Proceedin

Eighth International Workshop on Microprocessor Test and Verification
MTV 2007

5-6 December 2007
Austin, Texas, USA

Proceedings

Eighth International Workshop on Microprocessor Test and Verification
MTV 2007

5-6 December 2007
Austin, Texas, USA

Sponsored by
IEEE Computer Society Test Technology Technical Council
IBM
Freescale Semiconductor Inc.

Los Alamitos, California

Washington • Tokyo

Copyright © 2008 by The Institute of Electrical and Electronics Engineers, Inc.

All rights reserved.

Copyright and Reprint Permissions: Abstracting is permitted with credit to the source. Libraries may photocopy beyond the limits of US copyright law, for private use of patrons, those articles in this volume that carry a code at the bottom of the first page, provided that the per-copy fee indicated in the code is paid through the Copyright Clearance Center, 222 Rosewood Drive, Danvers, MA 01923.

Other copying, reprint, or republication requests should be addressed to: IEEE Copyrights Manager, IEEE Service Center, 445 Hoes Lane, P.O. Box 133, Piscataway, NJ 08855-1331.

The papers in this book comprise the proceedings of the meeting mentioned on the cover and title page. They reflect the authors' opinions and, in the interests of timely dissemination, are published as presented and without change. Their inclusion in this publication does not necessarily constitute endorsement by the editors, the IEEE Computer Society, or the Institute of Electrical and Electronics Engineers, Inc.

IEEE Computer Society Order Number P3241
BMS Part Number CFP07MTV-PRT
ISBN 978-0-7695-3241-7
ISSN Number 1550-4093

Additional copies may be ordered from:

IEEE Computer Society	IEEE Service Center	IEEE Computer Society
Customer Service Center	445 Hoes Lane	Asia/Pacific Office
10662 Los Vaqueros Circle	P.O. Box 1331	Watanabe Bldg., 1-4-2
P.O. Box 3014	Piscataway, NJ 08855-1331	Minami-Aoyama
Los Alamitos, CA 90720-1314	Tel: + 1 732 981 0060	Minato-ku, Tokyo 107-0062
Tel: + 1 800 272 6657	Fax: + 1 732 981 9667	JAPAN
Fax: + 1 714 821 4641	http://shop.ieee.org/store/	Tel: + 81 3 3408 3118
http://computer.org/cspress	customer-service@ieee.org	Fax: + 81 3 3408 3553
csbooks@computer.org		tokyo.ofc@computer.org

Individual paper REPRINTS may be ordered at: <reprints@computer.org>

Editorial production by Patrick Kellenberger
Cover art production by Joe Daigle/Studio Productions
Printed in the United States of America by Applied Digital Imaging

IEEE Computer Society
Conference Publishing Services (CPS)
http://www.computer.org/cps

Eighth International Workshop on Microprocessor Test and Verification

MTV 2007

Table of Contents

Preface vii
Acknowledgment viii
Committees ix

Power Analysis

Assertion-Based Modal Power Estimation 3
    Sumit Ahuja, Deepak A. Mathaikutty, Sandeep Shukla, and Ajit Dingankar
Early Models for System-Level Power Estimation 8
    Dam Sunwoo, Hassan Al-Sukhni, Jim Holt, and Derek Chiou
Reduction of Power Dissipation during Scan Testing by Test Vector Ordering 15
    Wang-Dauh Tseng and Lung-Jen Lee

Formal Methods

Mechanized Certification of Secure Hardware Designs 25
    Sandip Ray and Warren A. Hunt Jr.
Application of Lifting in Partial Design Analysis 33
    Marc Herbstritt, Vanessa Struve, and Bernd Becker
Model Checking Bluespec Specified Hardware Designs 39
    Gaurav Singh and Sandeep K. Shukla
Runtime Verification of k-Mutual Exclusion for SoCs 44
    Selma Ikiz and Alper Sen
A Scalable Symbolic Simulator for Verilog RTL 51
    Sasidhar Sunkari, Supratik Chakraborty, Vivekananda Vedula,
    and Kailasnath Maneparambil

System Level Validation and Test

Top Level SOC Interconnectivity Verification Using Formal Techniques 63
   Subir K. Roy

On Automatic Test Block Generation for Peripheral Testing in SoCs via Dynamic
FSMs Extraction 71
   D. Ravotto, E. Sanchez, M. Schillaci, M. Sonza Reorda, and G. Squillero

Automotive Microcontroller End-of-Line Test via Software-Based Methodologies 77
   W. Di Palma, D. Ravotto, E. Sanchez, M. Schillaci, M. Sonza Reorda, and G. Squillero

Functional Validation and ATPG

Intel® First Ever Converged Core Functional Validation Experience: Methodologies,
Challenges, Results and Learning 85
   Tommy Bojan, Igor Frumkin, and Robert Mauri

Chico: An On-chip Hardware Checker for Pipeline Control Logic 91
   Andrew DeOrio, Adam Bauserman, and Valeria Bertacco

A CLP-Based Functional ATPG for Extended FSMs 98
   Franco Fummi, Cristina Marconcini, Graziano Pravadelli, and Ian G. Harris

AMS Verification

Application of Automated Model Generation Techniques to Analog/Mixed-Signal
Circuits 109
   Scott Little, Alper Sen, and Chris Myers

Functional Modeling and Testbenches

An ADL for Functional Specification of IA32 119
   Wei Qin, Asa Ben-Tzur, and Boris Gutkovich

Automatic Testbench Generation for Rearchitected Designs 128
   Mark Nodine

Author Index 137

Preface
MTV 2007

The papers presented in this book have been revised from the original submissions to the Seventh IEEE International workshop on Microprocessor Test and Verification (MTV), held in Austin, Texas in December 2007 and sponsored by IEEE Computer Society Test Technology Technical Council (TTTC).

The topic area, applications of verification, validation and test to complex electronic circuits at all levels, has blossomed considerably since the first workshop was held in 1999. The scope of the workshop has expanded beyond just microprocessors to include all types of complex integrated circuits and Systems-on-Chip (SOCs). The 2007 workshop was certainly the most successful one of the series so far, and we would like to thank all participants who contributed to this event.

High level functional verification remains a key challenge facing designers of complex SOCs and microprocessors. This is reflected in large number of papers in this year's MTV. These papers discuss issues related to power analysis, formal methods, system level validation and test, functional validation and ATPG, analog & mixed signal verification and functional modeling and tests-benches. The boundaries of IP and SoC verification is an area that has evolved significantly in recent years and is being widely debated due to the blossoming of a wide array of related test and verification techniques. This year's MTV had a panel with a related topic – "IP vs. SoC verification: where one ends and the other begins".

To encourage industrial experts to openly discuss the current practice, we did not request written papers for publication for every presentation. Hence, this proceeding includes only the selected papers contributed by the authors and presenters. We will continue to adopt this strategy in order to encourage industrial participants to share their experience and results via the MTV forum. Interested readers who want to learn more about the current industrial practice can consult the MTV web site http://mtv.ece.ucsb.edu/MTV/ for future events.

Magdy S. Abadir, General Chair
Li-C. Wang, Program Chair
Jay Bhadra, Program Co-chair

Acknowledgment
MTV 2007

Many people contributed to the success of MTV and to the publishing of this proceeding. We thank all the contributors for their sustained interest, support and cooperation. We are deeply indebted to all members of the organizing and program committees for their support to MTV events over the years.

We also would like to thank our sponsors – the Test Technology Technical Council and the IEEE Computer Society.

Many thanks are due to IBM and Freescale Semiconductor for their continued support to MTV and for all technical and monetary contributions.

Special thanks to Anne Shelton for helping with various logistics during MTV. We would also like to thank Patrick Kellenberger for helping with production issues related to the publication of this proceedings.

Committees
MTV 2007

General Chair
Magdy S. Abadir, Freescale

Program Chair
Li-C. Wang, University of California, Santa Barbara

Program Co-chair
Jay Bhadra, Freescale

Organizing Committee

Finance
Jennifer Dworak, Brown University

Publication
Alper Sen, Freescale

Panel
Al Crouch, Inovys

Publicity
Tao Feng, Cadence

Committee
Moshe Levinger, IBM - Israel

European/Canadian
Andreas Veneris, University. of Toronto

Program Committee
Jacob Abraham, University of Texas, Austin
Miron Abramovici, DAFCA
Hussain Al-Asaad, University of California, Davis
Tony Ambler, University of Texas, Austin

Eyal Bin, IBM - Haifa
Shawn Blanton, CMU
Melvin Breuer, USC
Ken Butler, TI
K.-T. (Tim) Cheng, University of California, Santa Barbara
Nick Dutt, University of California, Irvine
Sujit Dey, University of California, San Diego
Ajit Dingankar, Intel
Franco Fummi, Universita `di Verona
Mike Garcia, Freescale
Sandeep Gupta, USC
Ian Harris, University of Massachusetts
John Hayes, University of Michigan
Eric Hennenhofer, Obsidian, Inc.
Jim Holt, Freescale
Alan J. Hu, UBC, Canada
T. M. Mak, Intel
Anmol Mathur, Calypto
Hillel Miller, Freescale
Sankaran Menon, Intel
Ishwar Parukar, Sun
Carl Pixley, Synopsys, Inc.
Paolo Prinetto, Poli di Torino
WangQi Qiu, Pextra Corp
Nur Touba, University of Texas, Austin
Miroslav Velev
Vivekananda Vedula, Intel
Cheng-Wen Wu, National Tsing-Hua University
Paul R Zehr, Intel
Yervant Zorian, VirageLogic

Power Analysis
MTV 2007

Assertion-Based Modal Power Estimation

Sumit Ahuja*, Deepak A. Mathaikutty*, Sandeep Shukla* and, Ajit Dingankar[†]
* CESCA, Virginia Tech, Blacksburg, VA 24061
[†]Design Technology and Solutions, Intel Corporation, Folsom, CA 95630
{sahuja, mathaikutty, shukla}@vt.edu and ajit.dingankar@intel.com

Abstract

Embedded Systems are becoming complex day by day and their increasing demand with shorter time-to-market is forcing designers to migrate to Electronic System-Level (ESL). One of the biggest issues with such battery-operated electronics is the power consumption. Facilitating power-aware architectural exploration at ESL requires a fast and accurate system-level power analysis capability. Existing frameworks suffer either in terms of accuracy or from the long turn-around time associated with the lower-level power analysis techniques. In this paper, we propose a power estimation technique for a specific class of design that is modal in nature (operation modes). The technique is illustrated through existing system-level design, verification and power estimation frameworks. The technique discussed in this paper provides an overview on how assertions written for verifying the reachability of modes can be utilized to generate directed test cases, which are given as an input to a power estimation framework.

I. Introduction

Designing System-on-Chips (SoCs) is becoming very difficult because of the high demand for multi-function, low power ubiquitous and battery-driven electronics. Due to the increasing complexity and stringent time-to-market requirements, the industry is migrating to the next level of abstraction, the Electronic System Level (ESL), where they expect to get a handle on this difficulty. ESL is an emerging design methodology that focuses on higher abstractions and allows the designer to create virtual system prototypes, perform rapid architectural exploration and power optimizations before the detailed hardware implementation (RTL) is finalized.

Performing accurate and efficient power analysis as early as possible in the design flow is very important in creating a power optimized design and is essential for the successful adoption of ESL methodologies in design flows. Currently accurate power analysis at system-level relies on power analysis methodologies developed at the register transfer level (RTL) or gate-level. The major bottleneck for power estimation is logic simulation that mandates large analysis time, because of the huge amount of information that needs to be processed at these levels. Some of the newer system-level techniques perform analysis based on relative-power estimation [1] or probabilistic power estimation [2]. However, all of the techniques either suffer in terms of accuracy or from the long turn-around time. We believe these problems can be avoided for one of the class of designs, which we call as **modal designs**.

Modal designs operate in different modes and have several "key" signals by which their operation modes are determined. Such a design approach is commonly taken for battery-driven devices such as mobile phones, PDAs and laptops. For example, the modes of operation for a mobile phone can be divided into talking and non-talking; in laptops these can be divided into active, idle, sleep, etc. Some of the control intensive designs in real time embedded systems such as traffic light controller, Elevator, etc. can also be considered as modal designs. These designs can be divided into such modes although they are not designed keeping the intent of power management in mind. In such designs, system goes to particular state for performing any operation and that state can be considered as a mode of operation. The power consumption of these designs is dependent on the operation modes. Therefore, computing the switching activity and the power dissipated in each mode will help us determine the overall power consumption of these systems. Designing these modal systems at high-level also require maintaining these notions of modes and values, which are exploited to perform accurate and efficient power estimation.

System-level power analysis for modal designs require associating power numbers with each state/mode of operation as well as with each system transition, which results in a state change. The system-level simulation, which is almost 100X faster than logic simulation is partitioned dynamically or statically (employing the Value Change Dump (VCD)) into "how long the system remain in one mode?" and "how long the system takes to transition to a particular mode?" Each partition of the simulation duration and the power number associated with that corresponding system *state* or *transition* are employed in computing the overall power consumption of the modal design.

In this paper, we propose a system-level power estimation technique for modal designs. As our implementation strategy, we propose to utilize an existing system-level design framework [3] and RTL power estimation infrastructure [2]. In [3] system-level design is specified using the ESTEREL language [4]. The system-level simulation and verification is performed using a commercial tool called Esterel Studio (ES) [5]. The power numbers associated with each mode and transition are obtained by utilizing PowerTheater [6], which is an RTL power analysis tool. The inputs for PowerTheater are generated from system-level **assertions** derived from properties [3], which are employed in the verification of the modal design. Such a technique reuses RTL power analysis with system-level inputs so that the designer gets sufficiently accurate power estimates in a shorter turn-around time.

A. Main Contributions

The main contributions of this paper are as follows:
- A system-level modal power estimation technique utilizing assertions.
- A detailed discussion of our power analysis technique through a system-level framework, which integrates ESTEREL [4] for system-level modeling, Esterel Studio [5] for verification and high-level synthesis, and PowerTheater [6] for RTL power estimation.

B. Organization

In section II we compare our work with existing methodologies. Section III provides the necessary background of languages and tools that can be used for implementation and section IV discuss in detail about the proposed power estimation technique. Finally, we conclude with a brief summary and the future work.

II. RELATED WORK

An extensive amount of work has been done on power estimation and optimization at the RTL and lower levels of abstraction [7], [8], [9]. They provide results with good accuracy, but are time consuming and not suitable for ESL designers.

In the recent past, researchers have proposed few methodologies for system-level power estimation. Caldari et al. presented a relative-power analysis methodology [1] for system-level models of the micro-controller bus architecture (AMBA) and high-performance bus (AHB) from ARM. It relies on creating macro-models from the knowledge of the possible implementations. Similarly, Bansal et al. presented a framework in [10], which uses the power-models of the components available at the system-level simulation stage by observing them at run time. It selects the most suitable power-model for each component by making efficiency and accuracy trade-offs. None of these work target designs having different modes of operation. Negri et al. presented a power modeling and simulation flow utilizing state-chart formalism for communication protocols in [11]. They have demonstrated their approach on communication protocols while our approach targets embedded hardware, also their approach is mainly applicable to the designs utilizing state-chart formalism but our methodology does not require any specific modeling style.

Shin et al. have proposed a methodology [12] for power estimation of operation modes but their analysis is done at logic-level and proposes a way to create power models based on the switching frequencies. Our approach is different from their approach as we are targeting the modal designs for power estimation at system-level keeping good accuracy in mind.

Zhong et al. have proposed a power estimation methodology [13] for cycle-accurate functional description of the hardware model. Here the idea is to utilize the cycle-accurate functional simulation for power estimation, which is assumed to be faster than corresponding RTL/gate-level power estimation approach. They have also emphasized the need of power estimation at higher-level of abstraction by utilizing state sampling techniques for finding out the different states that design can reach. Our approach is very different from their approach, as we do not use power macro-model [14] for power estimation hence accuracy of their approach is dependent on quality of macro-models, we rely on accurate power estimation using assertions from RTL description of the design. In addition, if power macro-models are not available during initial design phase then their approach is not useful while our approach does not have any such prerequisite hence useful during design phase as well.

III. TOOLS AND LANGUAGES USED

A. ESTEREL Language

ESTEREL is an imperative synchronous programming language for modeling synchronous reactive systems [4], especially suited for control-dominated system. It has various constructs to express concurrency, communication and preemption, whereas data-handling follows from procedural languages such as C.

B. Esterel Studio

Esterel Studio (ES) [5] is a development platform for designing reactive systems, which integrates a GUI for design capture, a verification engine for design verification and code generators to automatically generate target-specific executables. The GUI enables functional modeling through the graphical state machine specification called SSM or the ESTEREL textual specification [4]. The next most important ingredient of ES is the formal verifier (esVerify), which is used to verify the designer's intent. It allows both assertion-based verification as well as sequential equivalence checking. Finally, ES performs multi-targeted code generation, which range over targets such as RTL (VHDL/Verilog) to ESL (high-level C).

C. PowerTheater

PowerTheater [6] is an RTL/gate-level power estimation tool, which provides good accuracy for RTL power estimation with respect to the corresponding gate-level and silicon implementation. PowerTheater (PT) accepts design description in Verilog, VHDL or mixed verilog and vhdl. Other input required for average power analysis is value change dump in vcd or fsdb format dumped from RTL simulation. It also requires power characterized libraries in .lib or .alf format for power analysis. Apart from doing average power estimation it also helps user to do the activity analysis for testbenches, probabilistic power estimation, time-based power estimation, etc. In this paper, we have used simulation based average and time-based power estimation features of the tool.

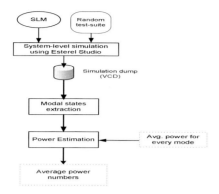

Fig. 1. Detailed view of system-level power estimation technique

IV. OUR POWER ESTIMATION TECHNIQUE

Fig. 1 shows the detailed view of our technique at system-level utilizing the average power associated with each mode and transition. Figure 2 provides the detail on how to obtain the average power number associated with each mode and transition. For applying our power estimation technique designer require a high-level designing and verification framework. We are using Esterel Studio for designing and verifying the system-level models (SLM). System-level models (SLM) are written in cycle-accurate transaction level (CATL) style in ESTEREL. From high-level simulation (performed in Esterel Studio) extract the total time spent in each state, transitions and the total number of transitions. We also propose to utilize the average power consumption in each mode at RTL (using PowerTheater) and calculate the power at system-level. The main advantage of our approach is that we do not require to do the power estimation at RTL for every testbench from the testsuite for the whole design. We utilize average power associated with each mode for estimation purpose at system-level.

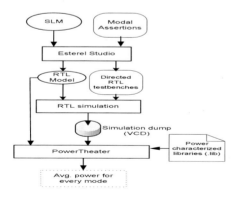

Fig. 2. Power estimation for each mode at RTL

Following are the steps that needs to be followed for power estimation in this framework (Fig. 1) :

1) Simulate the System Level Model (SLM) written in ESTEREL using Esterel Studio and store the Value Change Dump (VCD) for further analysis.

2) Extract different modes and transitions, total number of transitions, total time spent in each mode and transitions from the simulation dump as discussed in section IV-B.
3) Once we know the time spent in each state and transition, estimate the power of each state and transition then calculate the average power of the design using system-level simulation as discussed in section IV-C. This utilizes the average power number for different modes using assertions as shown in Fig. 2. Here are the steps to obtain the average power number for each mode/transition (Fig. 2):
   a) Create the cycle-accurate RTL model corresponding to the system-level cycle-accurate description in ESTEREL using the high-level synthesis engine provided by Esterel Studio (ES). ES is also used to convert the modal assertions (as discussed in section IV-A) to directed RTL testbenches.
   b) Once the RTL model and testbenches are ready, simulate the RTL model using RTL simulator (VCS [15]) and generate the simulation dump.
   c) The dump generated at RTL, power characterized libraries and all the required inputs are then passed to the RTL power estimation tool (PowerTheater). The average power numbers calculated at this stage are then utilized for power estimation during system-level simulation. This is one time process for as these numbers can be utilized for multiple system-level simulations for power estimation.

A. Assertions for finding out particular mode of design

In a typical design flow, properties are written for verifying the behavior of the design. In a design flow where design can work in different modes, one can write assertions to excite the different modes (states) of a design and to trigger a mode change (transition). We assume that the properties are expressed as assertions and enforced during the verification stage of the design. Such that they are available during the power estimation stage, if not then we write the assertions for testing the reachability of states and transitions. We illustrate some reachability assertions written for a simplistic modal design shown in Listing 1. Note that the design is specified as a pseudo-code and not in any specific language.

Listing 1. A simple modal design

```
1
2  module XYZ
3  begin
4    bool c1, c2;
5    string state;      // Possible values A, B, C, D
6
7    if(c1 == 0 && c2 == 0) then
8      state = "A";
9    else
10     if(c1 == 1 && c2 == 0) then
11       state = "B";
12     else
13       if(c1 == 0 && c2 == 1) then
14         state = "C"
15       else
16         state = "D"
17       end if
18     end if
19   end if
20 end
```

The design shown in Listing 1 has a variable state that is used to capture the mode of the system, which basically has four different values. The Boolean control variables `c1` and `c2` are used to trigger the required mode change. Assertions written for state reachability properties are shown in Listing 2.

Listing 2. Assertions to verify the reachability of states
```
// Property specifies the condition that reaches state A

assert always ( (c1=0 && c2=0) -> (state='A') );

// Property specifies the condition that reaches state D

assert always ( (c1=1 && c2=1) -> (state='D') );
```

An assertion that triggers the system to change from one state to another is shown in Listing 3.

Listing 3. Assertions to verify the reachability of a transition
```
// Property specifies the conditions that causes a transition from state D to C

assert always ( (c1=1 && c2=1 && next (c1=0)) -> ((state='D') && next(state='C')) );
```

The assertions are utilized for creating directed test cases such that each testbench either drives the system to a mode or causes the system to change its mode of operation. We utilize Esterel to express the assertions and Esterel Studio to generate the corresponding testbenches. Note that these directed testbenches are given as input to PowerTheater to estimate the expected power in the various operating modes of the system. These power numbers are reused at system-level to compute the overall power consumption of the system.

B. Extraction of modes from the simulation dump

As discussed earlier at high-level there are control signals associated with each of the mode, any change in the value of these control signals is dumped in the simulation-dump. Signals for these modes can have value depending upon its type e.g. if there is a boolean associated with mode then we check when the signal is true, else if it is short then we check the exact value of the variable associated with that mode. We extract all the possible time-stamps for which the mode signal remains in the expected value and how many number of times design goes to particular mode for calculating the total time spent in each mode. This can easily be done as VCD contains all the information related to the value of the mode. Similarly we calculate the total number of transitions (from one mode to another) occurring in the dump. This knowledge is then used in doing system-level power estimation of the design.

C. System-level power estimation

To calculate the system level power we first try to establish a relationship using energy spent in each mode during the full simulation duration and then we establish a relationship for the power at each state. Lets say that there are n different states and m different transitions in the design. Energy spent for the state i can be represented as E_i and total time spent in the state i can be represented as t_i. Similarly E_j is the energy associated with transition j, t_j is the total time spent on the transition j. If E_{total} is the total energy consumed in the design then we can establish the following:

$$E_{total} = \sum_{i=1}^{n} E_i + \sum_{j=1}^{m} E_j \quad (1)$$

If total average power is P_{total} and total simulation duration is T then from Equation 1 we can establish the following:

$$P_{total} * T = \sum_{i=1}^{n} P_i * t_i + \sum_{j=1}^{m} P_j * t_j \quad (2)$$

If a_i, a_j is represented as a fraction of total simulation duration spent in state i and transition j, we establish a relationship between a_i, a_j, t_i, t_j and T as shown in the Equation 3, 4.

$$a_i = t_i/T \quad (3)$$

$$a_j = t_j/T \quad (4)$$

In Equation 2 if we divide both the sides by T then we will get the following:

$$P_{total} = \sum_{i=1}^{n} P_i * a_i + \sum_{j=1}^{m} P_j * a_j \quad (5)$$

Which can further be simplified as:

$$P_{total} = \sum_{i=1}^{n} P_{Hi} + \sum_{j=1}^{m} P_{Hj} \quad (6)$$

In Equation 6, P_{Hi} and P_{Hj} represents the component of expected average power spent in each state and transition respectively. Equation 5 and 6 establishes the relationship between average power of each state, transition and total power calculated at high-level. In our approach P_i and P_j is calculated at RTL by utilizing the assertions written during verification stage (as discussed in section IV-A) and a_i and a_j is calculated as discussed in section IV-B. Finally P_{Hi}, P_{Hj} is calculated from the values of P_i, P_j, a_i and a_j.

V. CONCLUSIONS

In this paper, we propose a system-level power analysis technique that estimates power consumption of modal designs. We have discussed our implementation strategy utilizing Esterel Studio for high-level modeling & simulation as well as verification and PowerTheater for RTL power analysis. Our initial results on a sample modal design shows 98.6% accuracy when compared to a simulation-based RTL power estimation technique [16]. Our technique employs system-level simulation and power numbers borrowed from an RTL estimation tool driven by directed testbenches, to quickly compute the power consumption of a modal system design. Furthermore, we can also utilize our technique to find the MIN-MAX bound on the power consumption of the design. Methodology discussed in this paper is also applicable to an estimation flow, where the RTL implementation is developed separately as opposed to synthesizing it from the system-level design.

REFERENCES

[1] M. Caldari, M. Conti, M. Coppola, P. Crippa, S. Orcioni, L. Pieralisi, and C. Turchetti, "System-level power analysis methodology applied to the AMBA AHB Bus," in *Proceedings of the conference on Design, Automation and Test in Europe (DATE)*, 2003.

[2] S. Ahuja, D. Mathaikutty, G. Singh, S. Shukla and A. Dingankar, "SPAF: A System-level Power Analysis Framework," http://fermat.ece.vt.edu/Publications/pubs/techrep/techrep0707.pdf.

[3] D. Mathaikutty, S. Ahuja, A. Dingankar, S. Shukla, "Model-driven test generation for system level validation," IEEE International High Level Design Validation and Test Workshop (HLDVT07).

[4] G. Berry, "The foundations of ESTEREL," *Proof, Language and Interaction: Essays in Honour of Robin Milner*, pp. 425–454, 2000.

[5] ESTEREL Technologies, "Esterel Studio," http://www.esterel-technologies.com/products/esterel-studio/.

[6] Sequence Design Inc., "RTL power management," http://sequencedesign.com/solutions/powertheater.php.

[7] J. Frenkil, "Tools and methodologies for low power design," in *DAC '97: Proceedings of the 34th annual conference on Design automation*. New York, NY, USA: ACM Press, 1997, pp. 76–81.

[8] A. Raghunathan, N. K. Jha, and S. Dey, *High-Level Power Analysis and Optimization*. Norwell, MA, USA: Kluwer Academic Publishers, 1998.

[9] E. Macii, M. Pedram, and F. Somenzi, "High-level power modeling, estimation, and optimization," in *Proceedings of the Design automation conference*, 1997.

[10] N. Bansal, K. Lahiri, A. Raghunathan, and S. T. Chakradhar, "Power monitors: A framework for system-level power estimation using heterogeneous power models," in *Proceedings of the 18th International Conference on VLSI Design (VLSID'05)*, 2005.

[11] L. Negri and A. Chiarini, "Power simulation of communication protocols with statec," in *Applications of Specification and Design Languages for SoCs*, 2006.

[12] H. Shin and C. Lee, "Operation mode based high-level switching activity analysis for power estimation of digital circuits," in *IEICE transactions on Communications, E90-B(7):1826-1834*, 2007.

[13] L. Zhong, S. Ravi, A. Raghunathan, and N. K. Jha, "Power estimation for cycle-accurate functional descriptions of hardware," in *Proceedings of IEEE/ACM International Conference on Computer-Aided Design, San Jose, CA, Nov. 2004, p. 668*, 2004.

[14] N. R. Potlapally, A. Raghunathan, G. Lakshminarayana, M. Hsiao, and S. T. Chakradhar, "Accurate power macro-modeling techniques for complex rtl components," in *Proceedings of International Conference on VLSI Design Bangalore, India, Jan. 2001, p. 235*, 2001.

[15] Synopsys Inc., "VCS Comprehensive RTL Verification Solution," http://www.synopsys.com/vcs/.

[16] S. Ahuja, D. Mathaikutty, S. Shukla and A. Dingankar, "Utilizing Assertions for Modal Power Estimation," http://fermat.ece.vt.edu/Publications/pubs/techrep/techrep0713.pdf.

Early Models for System-level Power Estimation

Dam Sunwoo[1], Hassan Al-Sukhni[2], Jim Holt[2] and Derek Chiou[1]
[1]The University of Texas at Austin, [2]Freescale Semiconductor, Inc.
{sunwoo, derek}@ece.utexas.edu, {hassan.alsukhni, jim.holt}@freescale.com

Abstract

Power estimation and verification have become important aspects of System-on-Chip (SoC) design flows. However, rapid and effective power modeling and estimation technologies for complex SoC designs are not widely available. As a result, many SoC design teams focus the bulk of their efforts on using detailed low-level models to verify power consumption. While such models can accurately estimate power metrics for a given design, they suffer from two significant limitations: (1) they are only available late in the design cycle, after many architectural features have already been decided, and (2) they are so detailed that they impose severe limitations on the size and number of workloads that can be evaluated. While these methods are useful for power verification, architects require information much earlier in the design cycle, and are therefore often limited to estimating power using spreadsheets where the expected power dissipation of each module is summed up to predict total power. As the model becomes more refined, the frequency that each module is exercised may be added as an additional parameter to further increase the accuracy.

Current spreadsheets, however, rely on aggregate instruction counts and do not incorporate either time or input data and thus have inherent inaccuracies. Our strategy for early power estimation relies on (i) measurements from real silicon, (ii) models built from those measurements models that predict power consumption for a variety of processor micro-architectural structures and (iii) FPGA-based implementations of those models integrated with an FPGA-based performance simulator/emulator. The models will be designed specifically to be implemented within FPGAs. The intention is to integrate the power models with FPGA-based full-system, functional and performance simulators/emulators that will provide timing and functional information including data values. The long term goal is to provide relative power accuracy and power trends useful to architects during the architectural phase of a project, rather than precise power numbers that would require far more information than is available at that time. By implementing the power models in an FPGA and driving those power models with a system simulator/emulator that can feed the power models real data transitions generated by real software running on top of real operating systems, we hope to both improve the quality of early stage power estimation and improve power simulation performance.

1 Introduction

Power consumption is one of the key design constraints in microprocessor design, both in low-end embedded designs represented by portable handheld devices and in high-end high-performance designs such as servers and workstations. Chip designers can no longer pursue a performance-centric, power-unaware approach as the increased temperature and power consumption impact the overall system reliability.

Power optimization can be applied at several levels of the system, ranging from the lower level circuits to the upper architectural level. At the circuit level, techniques such as voltage scaling and clock gating could be applied. At the architectural level, idle components could be switched off or be switched to a low power mode. Many of these design decisions should be made at the early design phase to ensure timely delivery of the product. The dilemma is that the level of detail available at the early design phase is generally not deep enough to make accurate estimates of power consumption.

Spreadsheets from aggregate instruction counts are often used to predict power at the early design stage. These spreadsheets, however, do not incorporate either time or input data, thus leading to inherent inaccuracies.

Several architectural power estimation tools have been introduced[7, 25, 18] and have been used for various studies. Most of these tools are based on software performance simulators, such as SimpleScalar[3]. Although, these software simulators are widely adopted for their flexibility, they often suffer in simulation performance and accuracy. The accuracy of the performance simulation directly affects that of the power estimation. The speed of the simulation also affects the length of the simulation, which is often critical when doing temperature-sensitive studies.

FPGA-Accelerated Simulation Technologies(FAST)[9, 10, 12, 11] introduced a novel simulation methodology that enables fast and accurate simulation. The key idea is to partition a simulator into functional and timing models, and run the timing part in FPGAs to exploit parallelism. Such simulators can run orders of magnitude faster than traditional software simulators while maintaining very high accuracy. FAST simulators model a full system and are able to boot and run unmodified operating systems and applications.

In this paper, we propose power models using FAST simulators. Traditional power models that were used with software simulators can directly benefit from the speed and accuracy gained by FAST simulators. We also propose new power models that could run efficiently within the FPGA to enable even faster power estimation. Such a simulator will not only be useful to hardware designers, but also to software developers that can experiment various software implementations resulting in different amount of power consumption.

The rest of the paper is structured as follows. In section 2, we briefly overview the basics of FAST simulators. Section 3 proposes methodologies to write power models for FAST simulators. Section 4 discusses current status of the implementation. We then review the related work and conclude.

2 FAST Simulators

2.1 Overview

FAST simulators[9, 10, 12, 11] are the result of an effort to greatly enhance the performance of simulators without sacrificing any other features (accuracy, flexibility, etc.)

The main idea is to partition the simulator into a Functional Model(FM) and a Timing Model(TM) and to run each model on a platform that runs the model efficiently. The FM emulates the functional aspect of a system, including the behavior of the ISA and peripherals. General purpose processors are naturally good at executing instructions sequentially and this makes them a good platform to run the FM on.

The TM, on the other hand, models the timing aspect of the system. The timing of the system heavily depends on the structure of the components in hardware. Most modern computer hardwares employ several mechanisms to exploit more parallelism to achieve higher performance such as pipelining, out-of-order execution and superscalar processors. If we model such complex parallel structures using sequential programming languages, it is difficult to achieve high performance in simulation since they tend to serialize all the parallel execution. Modeling such structures in hardware, however, would allow us to enhance simulation

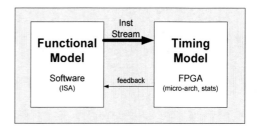

Figure 1. A high-level view of a FAST simulator

performance greatly. FPGAs are very well suited for such tasks because of the implicit parallelism of the hardware.

Thus, in FAST simulators, the TM is modeled in an FPGA. FPGAs enable us to try out different hardware designs quickly and efficiently. A high level view of a FAST simulator is shown in Figure 1. The FM is implemented in software running on a general purpose processor in the first prototype, although it could be put in hardware to further improve performance.

FAST simulators are able to tolerate the communication overhead between the FM and the TM since the two models are designed to run as loosely coupled as possible. The FM executes an instruction stream and sends traces containing information of the instructions to the TM. The FM, however, does not need to wait for the TM to process those traces before moving on the next instruction. As it has no notion of timing, the FM executes one instruction after another, always knowing the correct result of the previous instruction.

In this case, a problem occurs if the target architecture employs any sort of speculation, such as branch prediction. On a mispredicted branch, the processor may see wrong-path instructions filling up the pipeline, taking up resource and polluting the caches before the branch is resolved. This cannot be modeled correctly unless the FM provides the TM of the wrongpath instructions. As soon as the front end of the TM detects a branch is mispredicted, it notifies the FM. The FM, through checkpointing and rollback capabilities, rolls back to a previous checkpoint and proceeds down the desired wrongpath. After the mispredicted branch is resolved, the FM is notified again to continue down the right-path. Although this seems costly, the fairly high accuracy of modern branch predictors reduces the need for going through these rollbacks. In other words, the performance of a FAST simulator depends on the performance of the target architecture it is modelling.

Handling such speculation mechanisms, along with some back-pressuring, is the only feedback necessary from the TM. Such decoupling allows FAST simulators to efficiently tolerate the high communication cost with the FP-

GAs.

Writing modules for the TM is a very simple task. For most of the components in a target processor, the datapath can be omitted since they do not affect timing. For instance, caches can be modeled without data stores and ALUs do not need to actually compute any values. This greatly reduces the resource used by the TM and makes them efficient. Modules are allowed to take multiple FPGA host cycles to model a single target cycle. An eight-way set-associative structure may be modeled using an FSM that simply loops through up to eight host cycles, instead of trying to squeeze the operation in to a single clock cycle. This reduces the complexity of the TM and enables the TM to easily run at high frequencies. FAST connectors[12] are in charge of connecting up modules and synchronizing them behind the scene.

2.2 FAST Status

The first FAST simulator prototype was implemented based on the x86 ISA modeling a generic out-of-order superscalar processor. A detailed description of this prototype can be found in [11]. The FM was developed based on QEMU[4], an open-source emulator that employs dynamic compilation to achieve very high performance. QEMU was heavily modified to support trace generation and checkpointing/rollback capabilities. The TM was written in Bluespec[5], a high-level hardware description language that provides powerful features such as parameterizing interfaces at instantiation time. The prototype is implemented on a DRC Computer platform[14]. This machine has a dual-socket motherboard, where one socket is occupied by by an AMD Opteron 275 and the other by a Xilinx Virtex4 LX200 FPGA. The two chips communicate over HyperTransport. The FPGA module is also able to directly access several memory systems including the DRAM modules on the motherboard.

The first prototype is capable of booting unmodified operating systems such as Windows XP or Linux and running applications on top of them. The simulation performance ranges from 1 to 3 MIPS depending on the application running.

3 System Level Power Estimation using FAST

3.1 Benefits of using FAST simulators

So far, we have discussed how FAST simulators achieve high performance while maintaining accuracy. Since power consumption, especially dynamic power consumption, is heavily dependent upon nodes switching values, its estimation cannot be accurate without a very accurate model of the hardware structure and the switching timing.

As FAST simulators are capable of faithful modeling of the hardware, they could serve as an excellent candidate for estimating power consumption as well. The potential speed of the performance/power simulation may serve extremely useful not only to architects exploring design spaces, but also to software developers who are interested in the amount of power consumed by their program. Conventional power estimation tools are practically unusable to software developers due to their slow speed.

One could also imagine exploring temperature-aware policies at the microarchitectural level using FAST power simulation. Hotspot[23] originally explored architecture-level thermal management. Since temperature changes take place quite slowly (taking at least 100K cycles to rise by $0.1°C$,) a faster simulation would open up possibilities for exploring policies of longer terms.

FAST simulators also model speculation mechanisms very accurately, which is very important in terms of power consumption as well. Some software simulators assume constant mispeculation penalties that could lead to inaccurate evaluation of design decisions. FAST simulators are capable of getting more accurate results and will assist the user in making the correct decision.

3.2 Approach

To fulfill this objective, we propose to implement a FAST power simulator for an existing core. Although we will be modeling more aggressive cores, we start by implementing an accurate model of the Freescale e200z6 core[2]. The e200z6 core is based on the Power ISA[22] and has a single-issue, seven-stage pipeline. It also provides MMU, unified L1 cache, FPU and vector processing capabilities.

The e200z6 core is a relatively simple processor, yet contains many architectural components that may have different power consumption characteristics. The MPC5554 high performance microcontroller, which includes the e200z6 core, is also available packaged as an evaluation board[1] for development purposes. This board gives us the ability to easily validate and calibrate our simulator against actual implementations on silicon. More details on this will be presented in the following subsection.

3.3 Validation against Real Hardware

Once the initial port of FAST to the e200z6 core is finished, we plan to extensively calibrate the simulator against the actual hardware implementation using the MPC5554 evaluation board. The e200z6, as is the case with most modern microprocessors, provides a set of performance measurement counters that could be used to tune various components of the core. The TM of the e200z6 core, along with the memory hierarchy and some key peripherals could be tuned to be highly accurate.

We also plan to develop the ability to measure power consumption of the e200z6. Since the MPC5554 does not have separate power pins for the core, micro-benchmarks that are specifically crafted to stress different components and only those components will be written to assist the measurement of power consumption. For example, a series of add instructions could be used to stress the ALU, while a series of load/store instructions would characterize the cache. Of course, none of these instructions will utilize only the ALU or only the cache as all instructions still flow through the entire pipeline. Careful analyses would allow us to extract the desired information (For example, comparing against a series of NOP instructions.) These experiments will be similar to those conducted in [17] and [13].

3.4 Power Modeling Methodology

We are planning to take a three-phase approach to develop and refine our power model for FAST simulators.

3.4.1 Duplication of the Spreadsheet Approach

Initially, we will try to duplicate the spreadsheet approach that most current architecture-level power estimation methods take. FAST simulators will generate statistics of when each component is used and use that information to feed an equation that will compute the estimated power. The estimates will then be validated against actual measurements taken from the MPC5554 evaluation board. These results will also serve as a baseline comparison point in terms of accuracy for the new power models to be developed in the next phase.

3.4.2 Invent, Develop and Refine Initial Power Prediction Models

In this phase, we will develop novel power model modules that are intended to improve modeling accuracy during the architectural phase. The primary goal of this phase is to develop such modules that will efficiently fit in the FPGA.

Sensitivity studies will be performed to determine what parameters have the most impact on power. For example, it is likely that inputs that result in longer carry chains will consume more power in the ALU. In a floating point unit, the similarity of the exponents of the arguments will have a significant effect on power consumption. Also, a cache miss clearly has an impact on power, but might have less of an impact if there is already a cache miss outstanding.

As a power modeling experiment, the power characterization for a 6-stage floating point unit (FPU) was studied

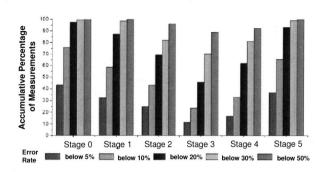

Figure 2. Accuracy of power models for each pipeline stage of a research FPU

using floating-point add instructions. We carried out detailed cycle-by-cycle gate-level power simulations for the FPU and extracted power models based on the simulation results using regression analysis. The purpose of the study was to identify correlations between the number of switching bits at the input of the FPU and its dynamic energy consumption. The hamming distance between two consecutive inputs to the FPU is defined as the number of bits that switch from 0 in the first input to 1 in the subsequent input.

From the experiments, we found out that the FPU power model is a complicated model of many parameters, and the hamming distance alone is not sufficient for estimating the FPU power. Based on the experiments, we found out that for the add instruction, the following parameters affect the dynamic power consumption of the FPU:

- Hamming distance between consecutive operands.
- Hamming distance between consecutive results.
- Alignment shift of operands.
- Normalization shift operations.
- Sign bits of the two operands.

Figure 2 illustrates the accuracy of the developed models for each stage of the FPU pipeline. For each stage the Figure depicts five bars. Each bar represents the accumulative percentage of power measurements that agree with the model within the specified error rate. The figure indicates that the prediction accuracy of the power model differs significantly for the different pipeline stages. For example, the estimated power consumption from the regression model for Stage 0 was within 20% error range of the measured power for about 95% of the samples. On the other hand, less than 50% of the samples were within the same error range for Stage 3.

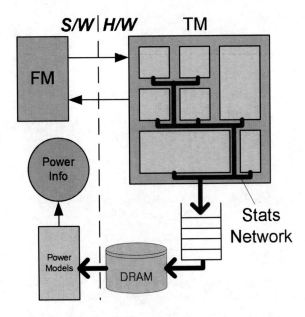

Figure 3. Power estimation by post-processing statistics

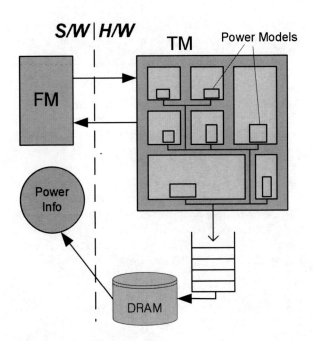

Figure 4. Power estimation using on-FPGA power models

The experiments indicate that there might be other meaningful parameters for intermediate stages that need to be explored, in addition to the parameters above. In our experiments, we also found out that the dynamic energy consumption differs significantly for different inputs, and an average number is not enough to capture the power characteristics. Our regression models were good for some stages of the pipeline. While the models were not completely satisfactory for other stages, we believe that we will be able to generate better models eventually.

Once we have power models for individual modules, we need to plan how to structure the overall simulator and place the models. At first, we will take a post-processing approach as shown in Fig. 3. We are currently implementing a *statistic network* that will efficiently drive various statistics out of the TM without causing global routing congestion. The statistics will contain dynamic information, such as input data or activity factors, that is needed to calculate power consumption. These numbers will exit the TM through an *activity queue*. Note that the multiple enqueue operations for a target cycle could be handled efficiently since the enqueues could be spread across multiple host cycles without further slowing down the simulation. The statistics can be buffered up in larger memory adjacent to the FPGA chip before being streamed out (A detailed description of our development platform will be provided in section 4.) A conventional software power model of components could be used to post-process the information and generate an estimate of the power consumed.

This post-processing approach, however, may lead to lower simulation performance depending on the level of detail of the power model. It will be the bottleneck and slow down the overall simulation dramatically. Ideally, we would want power models sitting inside the FPGA, as shown in Fig. 4, and have the power consumption estimated at full-speed at run-time. Note that in Fig. 4, the statistic network and the arrows going off-FPGA through the DRAM are drawn with thinner lines. This represents the lower bandwidth required compared to the post-processing approach in Fig. 3. Now, the problem becomes how to design a power model that tracks power fairly accurately and fits in the FPGA efficiently.

Conventional power models either employ analytical models[7] or empirical models[25]. Empirical models would be more suited for the FPGAs using table lookups. Clustering techniques[21] have been studied to minimize the size of the lookup table as the size could grow exponentially with the input data width. We are working on more efficient heuristics to further reduce the table size without sacrificing accuracy. Power macro-modeling techniques using input statistics[15] with improved input vectors[20] can be used to generate lookup tables with reasonable sizes. Recent studies[19] employ hybrid models of analytical and empirical techniques to improve accuracy. Our FPGA-based models will also use such techniques.

Power models that are implemented within an FPGA will result in run-time power estimation that runs fast enough to be useful not only to architects but also to software developers.

3.4.3 Extrapolation to Future Processes

While the previous phase focused on predicting power consumption of existing hardware, the ability to predict power for next-generation hardware is also very important. In this phase, we will develop extrapolation techniques to do so. Because, the e200z6 core is implemented in at least two processes, we will be able to use real hardware to experimentally determine the differences in power consumption between processes and what factors most influence those differences.

4 Current Status

As mentioned in section 2.2, the FAST prototype is up and running, modeling an out-of-order superscalar x86 processor. Some porting work is necessary to support the new target model.

As we are targeting to model the e200z6 core, the FM obviously needs to support the Power ISA. This port is almost complete at the point of this writing. The TM should be written very quickly as the e200z6 is much simpler than the original prototype FAST TM. We also expect to see much higher performance than the original due to the simplicity of the core. The prototype FAST TM, however, was not validated or calibrated against any existing hardware, which will be done for the e200z6. This process may bring up interesting aspects or issues in designing a highly accurate simulator.

We are also experimenting with various power models, including power macromodeling techniques introduced in [15, 20]. We believe that these modeling techniques can generate fairly accurate power models with reasonably sized (less than 1024 entries per module) lookup tables.

5 Related Work

Power estimation techniques at the architectural level have been explored extensively over the last decade. Wattch[7] uses analytical models based on node capacitance estimates and aggregate activity factors to model the key components in a processor. SimplePower[25, 24] uses lookup tables generated from detailed circuit models. The lookup table sizes are reduced using clustering techniques in [21]. Wattch and SimplePower are both based on SimpleScalar[3]. PowerTimer[6] uses power models from macro-level power simulation data from prior processors along with newer circuit-simulation experiments to characterize new components.

A recent study[19] proposed using a hybrid method of analytical and empirical techniques by building analytical models based on basic blocks modeled with empirical data.

Kim et al.[18] derived very detailed circuit models that are then driven by Simplscalar. However, the simulation would be slow and always requires a detailed design, which is not suitable for early stage exploration.

SoftWatt[16] attempts to estimate power for a full system, using SimOS targeting a MIPS R10000 processor and some peripherals. However, it is unable to produce cycle-by-cycle estimates. Note that FAST simulators also model full systems and can potentially predict power for such complete systems.

Static power at the architectural level was studied in [8] and [26]. We also plan to account for leakage power in future implementations of the FAST power simulator.

6 Conclusion and Future Work

In this paper, we have proposed a novel power estimation methodology that can predict power consumption fairly accurately at the early stage of the design. These models will either run along with or run within FPGA parts that would accelerate the simulation performance. The resulting simulator will be based on actual hardware implementations and will be validated and calibrated to ensure accuracy.

Currently, the prototype FAST simulator is being ported to the target processor and FPGA-based power models are being defined. Issues with designing power models for various hardware components targeted for FPGAs will be addressed in future studies.

References

[1] MPC5554 Evaluation Board. http://www.freescale.com.

[2] Freescale's e200 Core Family Built on Power Architecture Technology. White Paper, 2007.

[3] T. Austin, E. Larson, and D. Ernst. SimpleScalar: An Infrastructure for Computer System Modeling. *IEEE Computer*, 35(2):59–67, Feb. 2002.

[4] F. Bellard. QEMU, a Fast and Portable Dynamic Translator. In *USENIX 2005 Annual Technical Conference, FREENIX Track*, pages 41–46, 2005.

[5] Bluespec webpage. http://www.bluespec.com.

[6] D. Brooks, P. Bose, V. Srinivasan, M. K. Gschwind, P. G. Emma, and M. G. Rosenfield. New methodology for early-state, microarchitecture-level power-performance analysis of microprocessors. *IBM Journal of Research and Development*, 47(5):653–670, 2003.

[7] D. Brooks, V. Tiwari, and M. Martonosi. Wattch: a framework for architectural-level power analysis and optimizations. *Proceedings of the 27th annual international symposium on Computer architecture*, pages 83–94, June 2000.

[8] J. A. Butts and G. S. Sohi. A static power model for architects. *Proceedings of the 33rd Annual ACM/IEEE International Symposium on Microarchitecture*, pages 191–201, December 2000.

[9] D. Chiou. FAST: FPGA-based Acceleration of Simulator Timing models. In *Proceedings of the first Workshop on Architecture Research using FPGA Platforms, held in conjunction with HPCA-11, San Francisco, CA*, Feb. 2005.

[10] D. Chiou, H. Sanjeliwala, D. Sunwoo, J. Z. Xu, and N. Patil. FPGA-based Fast, Cycle-Accurate, Full-System Simulators. In *Proceedings of the second Workshop on Architecture Research using FPGA Platforms, held in conjunction with HPCA-12, Austin, TX*, Feb. 2006.

[11] D. Chiou, D. Sunwoo, J. Kim, N. A. Patil, W. Reinhart, D. E. Johnson, J. Keefe, and H. Angepat. FPGA-Accelerated Simulation Technologies (FAST): Fast, Full-System, Cycle-Accurate Simulators. *Proceedings of the 40th Annual IEEE/ACM International Symposium on Microarchitecture*, December 2007.

[12] D. Chiou, D. Sunwoo, J. Kim, N. A. Patil, W. Reinhart, D. E. Johnson, and Z. Xu. The FAST Methodology for High-Speed SoC/Computer Simulation. *2007 International Conference on Computer-Aided Design (ICCAD'07)*, November 2007.

[13] G. Contreras and M. Martonosi. Power prediction for Intel XScale® processors using performance monitoring unit events. *Proceedings of the 2005 International Symposium on Low Power Electronics and Design*, pages 221–226, 2005.

[14] DRC Computer. http://www.drccomputer.com/.

[15] S. Gupta and F. Najm. Power modeling for high-level power estimation. *Very Large Scale Integration (VLSI) Systems, IEEE Transactions on*, 8(1):18–29, 2000.

[16] S. Gurumurthi, A. Sivasubramaniam, M. Irwin, N. Vijaykrishnan, M. Kandemir, T. Li, and L. John. Using complete machine simulation for software power estimation: the SoftWatt approach. *High-Performance Computer Architecture, 2002. Proceedings. Eighth International Symposium on*, pages 141–150, 2002.

[17] C. Isci and M. Martonosi. Runtime power monitoring in high-end processors: methodology and empirical data. *Microarchitecture, 2003. MICRO-36. Proceedings. 36th Annual IEEE/ACM International Symposium on*, pages 93–104, 2003.

[18] N. Kim, T. Kgil, V. Bertacco, T. Austin, and T. Mudge. Microarchitectural Power Modeling Techniques for Deep Sub-Micron Microprocessors. *Low Power Electronics and Design, 2004. ISLPED'04. Proceedings of the 2004 International Symposium on*, pages 212–217, 2004.

[19] X. Liang, K. Turgay, and D. Brooks. Architectural Power Models for SRAM and CAM Structures Based on Hybrid Analytical/Empirical Techniques. *2007 International Conference on Computer-Aided Design (ICCAD'07)*, November 2007.

[20] X. Liu and M. Papaefthymiou. A Markov chain sequence generator for power macromodeling. *Computer-Aided Design of Integrated Circuits and Systems, IEEE Transactions on*, 23(7):1048–1062, 2004.

[21] H. Mehta, R. M. Owens, and M. J. Irwin. Energy characterization based on clustering. *Proceedings of the 33rd annual conference on Design automation*, pages 702–707, June 1996.

[22] Power webpage. http://www.power.org.

[23] K. Skadron, M. Stan, W. Huang, S. Velusamy, K. Sankaranarayanan, and D. Tarjan. Temperature-aware microarchitecture. *Proceedings of the 30th Internation Symposium on Computer Architecture*, June 2003.

[24] N. Vijaykrishnan, M. Kandemir, M. J. Irwin, H. Kim, and W. Ye. Energy-driven integrated hardware-software optimizations using simplepower. *Proceedings of the 27th annual international symposium on Computer architecture*, pages 95–106, June 2000.

[25] W. Ye, N. Vijaykrishnan, M. Kandemir, and M. Irwin. The design and use of simplepower: a cycle-accurate energy estimation tool. *Proceedings of the 37th Conference on Design Automation*, pages 340–345, 2000.

[26] Y. Zhang, D. Parikh, K. Sankaranarayanan, K. Skadron, and M. Stan. Hotleakage: A temperature-aware model of subthreshold and gate leakage for architects. Technical Report CS-2003-05, University of Virginia, Department of Computer Science, 2003.

Eighth International Workshop on Microprocessor Test and Verification

Reduction of Power Dissipation during Scan Testing by Test Vector Ordering

Wang-Dauh Tseng[1][†] and Lung-Jen Lee[2]

[1]*Department of Computer Science & Engineering*
Yuan Ze University
Chung-li, Taiwan, 32003, ROC
[†]wdtseng@saturn.yzu.edu.tw

[2]*Department of Electronic Engineering*
National Army Academy
Chung-li, Taiwan 32092, ROC

Abstract

Test vector ordering is recognized as a simple and non-intrusive approach to assist test power reduction. Simulation based test vector ordering approach to minimize circuit transitions requires exhaustive simulation of each test vector pair. However, long simulation time makes this approach impractical for circuits with large test set. In this paper we present a calculation based approach to faster order test vectors to reduce test power for full scan sequential circuits. Most calculation approaches are for combinational circuits or for sequential circuits but only considering the portion of circuit derived from the primary inputs. The proposed approach exploits the dependencies between internal circuits and transitions at both the primary and state inputs. Experiments performed on the ISCAS 89 benchmark circuits show that the improvement efficiency of the proposed approach can achieve 91.55% and has better performance than the existing calculation based approaches.

1. Introduction

Low power electronics has become increasingly important with the advent of portable electronic devices. This has motivated designers to reduce power consumption in the circuit, during both normal operation and testing. Power consumption of digital systems is considerably higher in test mode than in normal mode. During system normal operation, low power consumption can be attributed to the significant correlation that exists between successive vectors applied to a given circuit, whereas in test mode this is not necessarily true. Elevated test power may cause logical error in a fault-free chip leading to an unnecessary loss of yield. Hence it is important to reduce power consumption during test application. Various techniques have been proposed to reduce power consumption during test application [1-7]. Because power consumption in CMOS circuits is proportional to the switching activity in the circuit, the majority of these techniques concentrate on reducing the power consumption by minimizing the switching activity. The technique for power minimization with no penalty in test area and performance is based on test vector ordering which modifies the order in which test vectors of a given test sequence are applied to the CUT [1-5]. The authors in [1] construct a complete weighted graph, called transition graph, in which each vertex represents a test vector and a weight assigned to each edge represents the number of transitions activated in the circuit due to the application of the test vector pair connecting to the edge. Logic and timing simulations are required to compute the weight assigned to each edge. There are totally $n(n-1)/2$ simulations required for the construction of the graph, where n is the number of test vectors. The general delay model is assumed during simulations. A greedy algorithm is then used to find a Hamiltonian path of minimum cost in the transition graph. The main problem in this approach rests in the time needed to construct the transition graph for circuits with a large number of test vectors. In order to reduce the graph construction time, the paper in [2] uses zero-delay model for logic and timing simulations. Although the simulation time is reduced, the number of simulations remains $n(n-1)/2$ and the time required to construct the transition graph is still high. The paper in [3] propose a fast simulation method which only take into account the expected switching activity at the primary inputs and at a very small set of internal lines of the CUT. The computational time for the construction of transition graph is reduced but the power reduction obtained in [3] is lower than that in [1] and [2]. To make it possible to apply test vector reordering to circuits with large number of test vectors, the authors in [4] employ the Hamming distance between test vectors rather than simulate transitions in the circuit to evaluate the power consumption. Although the Hamming distance approach is the most time-saving, it doesn't take into account the

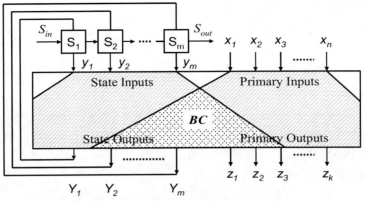

Figure 1. A full scan sequential circuit

topology of the circuit. The reduction in power consumption for the Hamming distance approach is usually significantly lower than that for the simulation based approaches [1-3]. The authors in [5] consider the structure of the CUT in the estimation of the weights in the transition graph thus providing better solutions in terms of amount of power saved. An induced activity function is proposed in this paper to measure the impact of a transition at a specified input on the switching activity of the CUT and is used as the weight of the Hamming distance. However, these calculation based approaches in [2-5] are only applicable to combinational circuits or full scan sequential circuits with specific scan cell design. In this paper we present a calculation based test vector ordering technique that reduces power consumption during test application for full scan sequential circuits. We exploit the dependencies between internal circuits and the transitions at scan cells as well as primary inputs. The idea behind the proposed approach is based on the following two observations. First, for a scan cell, the number of transitions caused by a test vector being scanned in/out depends not only on the transitions in the test vector/response but also on its position in the scan chain. Second, the impact of each circuit input on the switching activity in the internal circuit is different. If a transition at a circuit input of the CUT propagates to the internal circuit, it will result in a large number of unnecessary transitions. Depending on the circuit structure, the transitions at some circuit inputs may cause more transitions at internal circuit than those at other circuit inputs. Based on the two observations, weights on the transition graph for test vector ordering can be derived. In the following, two functions to compute the transition count of a scan cell in the scan chain and to measure the impact of a circuit input on switching activity in the internal circuit are developed, respectively. The rest of the paper is organized as follows. Section 2 provides the background on scan testing and some circuit definitions. In Section 3, an impact function is developed to measure the impact of transitions at a gate on the switching activity in the combinational part of the circuit. Section 4 calculates the transition count at each scan cell as a pair of test vectors is scanned. Section 5 presents the proposed approach. To validate the proposed approach, experimental results are given in Section 6. Section 7 concludes the paper.

2. Circuit definitions

Consider the full scan sequential circuit shown in Fig.1, comprised of a block of combinational circuit C and a set of m state elements. The primary inputs and outputs of the circuit are $x_1, x_2,..., x_n$ and $z_1, z_2,..., z_k$, respectively. The present state variables, $y_1, y_2,..., y_m$, constitute the state inputs of the combinational circuit. The next state variables, $Y_1, Y_2,..., Y_m$, constitute the state outputs of the combinational circuit. The *forward cone* of line l, $FC(l)$, is defined as the portion of a circuit whose signals are reachable by a forward trace of the circuit topology starting at l. The *primary forward cone* (*PFC*) is defined as the portion of the circuit which is the union of forward cones of all primary inputs. The change of values on primary inputs by applying consecutive test vectors causes switching activity in the primary forward cone. The *state forward cone* (*SFC*) is defined as the portion of a circuit which is the union of forward cones of all state inputs. The change of values on state inputs by shifting a scan vector causes switching activity in the state forward cone. The portion of the circuit defined as the intersection of primary forward cone and state forward cone is referred to as *blocking cone* (*BC*). Any gate in the blocking cone has at least one input in the path which starts from a primary input and one input in the path which starts from a state input. Most calculation based test vector ordering approaches to reducing test power is by lowering the transition density at the primary inputs. Very little work considers reducing the transition density at the state inputs; therefore, only the switching activity in the PFC of the CUT can be reduced by these approaches. In this paper, we try to lower the transition density at both the primary inputs and state inputs by ordering the test vectors such that the switching activity in the PFC and SFC of the CUT can be reduced.

3. Calculation of impact function

During scan cycle, filling in the scan chain with the state input part of a test vector requires shifting the bits one by one into each scan cell, thus creating increased switching activity in the scan cells. The rippling effect originating from a scan chain to the CUT results in a large number of unnecessary transitions at the circuit lines. Additionally, if a transition at a primary input of the CUT propagates to the internal circuit, it will subsequently cause transitions. Depending on the circuit structure, the transitions at some circuit inputs (primary /state inputs) of a CUT cause more transitions at internal lines than those at other circuit inputs. Therefore, reducing transitions at those circuit inputs that cause more transitions in the internal circuit will make greater reduction in switching activity. To measure the transition effect at a circuit input reflecting into the CUT, it is necessary to develop an approach to evaluate the impact on the transitive fanout of each circuit input. The authors in [7] propose a gain function for computing the weighted transition density, which provides an effective measure of the switching activity in logic circuit, for each primary input. In this section, we develop an impact function, which is modified from the gain function, to measure the transition of a state/primary input impact on the switching activity in the internal circuit.

Consider a CUT with m circuit inputs $x_1, x_2, ..., x_m$. The *signal probability* $sp(c)$ of a circuit line c is defined as the probability that c is set to 1:

$$sp(c) = Pr(c = 1) \quad (1)$$

Signal probability can be propagated through logic gates based on simple rules of probability and logic function of the gates. The *transition probability* of a circuit line c is the probability of the signal making a transition from one state to another at any time t and is denoted by $p_t(c)$. Under the assumption that the values applied to each circuit input are temporally independent, we can write:

$$p_t(c) = 2 \cdot sp(c) \cdot (1 - sp(c)) \quad (2)$$

Let $n_c(T)$ be the number of transitions at a circuit line c in a time interval of length T. The *transition density* at c, i.e. the number of transitions per second at c, is defined as:

$$D(c) = \lim_{T \to \infty} \frac{n_c(T)}{T} \quad (3)$$

Let f_i be a function that depends on circuit input x_i. The Boolean difference of f_i with respect to x_i is defined as follows:

$$\frac{\partial f_i}{\partial x_i} = f_i|_{x_i=1} \oplus f_i|_{x_i=0} \quad (4)$$

where $\oplus$ denotes the exclusive-or operation. The Boolean difference signifies the condition under which output f_i is sensitized to circuit input x_i. If the circuit inputs x_i, $i = 1, ..., m$, to the CUT are not spatially correlated, the transition density of a circuit line c can be defined in terms of the Boolean difference with respect to each circuit input, $\partial f_i / \partial x_i$, and the transition density of each circuit input, $D(x_i)$, as:

$$D(c) = \sum_{i=1}^{m} P\left(\frac{\partial f_i}{\partial x_i}\right) D(x_i) \quad (5)$$

The Boolean difference $\partial f_i / \partial x_i$ represents the condition for sensitizing circuit input x_i to output f_i as noted in Eq. (4). Therefore, $P(\partial f_i / \partial x_i)$ signifies the probability of sensitizing input x_i to output f_i, while $P(\partial f_i / \partial x_i)D(x_i)$ is the contribution of transitions at output f_i due to circuit input x_i only. Hence, the contribution of transitions at output f_i due to all the circuit inputs is obtained by taking the summation over all the circuit inputs of the CUT.

As shown in Eq. (5), the transition density of a circuit line c is the sum of the transitions at each circuit input that sensitize to line c. Hence, the portion of the transition density of line c due to the transition at a circuit input x_i, is given by

$$D_{x_i}(c) = P\left(\frac{\partial f_i}{\partial x_i}\right) D(x_i) \quad (6)$$

Similarly, the portion of the transition density of line c due to the transition at a specific line k, can be written by

$$D_k(c) = P\left(\frac{\partial g_i}{\partial k}\right) D(k) \quad (7)$$

where g_i is a function that depends on circuit line k. The sum of transition densities of lines in the forward cone of k, $FC(k)$, that can be attributed to the transitions at k is given by:

$$D_k = \sum_{\forall c \in FC(k)} D_k(c) \quad (8)$$

For the CMOS circuit technology, dynamic power due to the charging and discharging circuit capacitances is the dominant source of power consumption. Hence, the power dissipation in a circuit depends on the load capacitance of internal lines. However, lines connected to more gates are lines with higher parasitic capacitance. If two circuit lines have the same transition density, the one with higher fanout will consume more power than the other one with lower fanout. Load capacitance also depends on the type and the size of the device. For example, a 2-input NOR has more load capacitance than a 2-input NAND, and the load capacitance difference

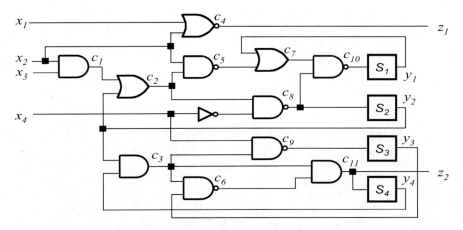

Figure 2. An example circuit

increases as the number of inputs increase. So, two factors, the fanout and device coefficient, are also considered in the impact function as the weights of the transition density. Although wire length also affects power consumption, for simplicity, it is not considered in the impact function. For each circuit line k the impact function IMP_k can be expressed as:

$$IMP_k = \sum_{\forall c \in FC(k)} F_c \alpha_c D_k(c) \quad (9)$$

where F_c and α_c are the fanout and device coefficient of circuit line c, respectively. The fanout of the lines is defined by circuit topology. The device coefficient can be obtained once the circuit has been synthesized. In this derivation, the Boolean difference of f_i with respect to c is derived from the signal probability. However the signal probability $P(\partial f_i / \partial sp_i)$ can easily be calculated using the similar procedure which is used to calculate detection probability [3].

To illustrate the calculation of impact function value, considers the circuit shown in Fig. 2. The primary inputs are $\{x_1, x_2, x_3, x_4\}$, $\{S_1, S_2, S_3, S_4\}$ are scan cells, $\{y_1, y_2, y_3, y_4\}$ are the state inputs, and $\{z_1, z_2\}$ are the primary outputs. The forward cone of primary input x_3 is composed of circuit lines c_1, c_2, c_5, c_7, c_8 and c_{10}. Table 1 shows the impact function value for primary input x_3 in Fig. 2. The first row shows the circuit lines in the forward cone of x_3. The second and third rows show the corresponding fanout and transition density of each circuit line in each column, respectively. The circuit lines and their fanouts included in the forward cone of x_3 can be obtained directly from the circuit structure. However the transition density for each circuit line can

Table 1. Calculation of impact function value for primary input x_3

$FC(x_3)$	c_1	c_2	c_5	c_7	c_8	c_{10}	IMP
Fanout	1	2	1	1	2	1	1.4
TD	1/4	1/4	1/8	1/16	1/16	3/32	

be calculated using Eqs. (4) and (6). Take circuit line c_5 for example. As shown in Fig. 2, the Boolean function f_{c5} for circuit line c_5 can be express as $\overline{(x_2 x_3 + y_2) x_2}$. By the definition in Eq. (4) the Boolean difference of f_{c5} with respect to x_3 can be expressed as $\frac{\partial f_{c5}}{\partial x_3} = \overline{(x_2 x_3 + y_2) x_2}\big|_{x_3=0} \oplus \overline{(x_2 x_3 + y_2) x_2}\big|_{x_3=1} = x_2 \overline{y_2}$.

Hence the signal probability for c_5, $f\left(\frac{\partial f_{c5}}{\partial x_3} = x_2 \overline{y_2}\right)$, is equal to 1/4. Assume that the signal probability for each bit in a test vector is 1/2. By Eq. (2) the transition probability for x_3 is equal to $2 \times 1/2 \times (1 - 1/2) = 1/2$. According to Eq. (6), the transition density for c_5 is equal to $1/4 \times 1/2 = 1/8$. Assuming that the device coefficient is 1 for all type of gates, the impact function value of x_3 can be calculated by Eq. (9) and is equal to $1 \times 1/2 + 2 \times 1/4 + 1 \times 1/8 + 1 \times 1/16 + 2 \times 1/16 + 1 \times 3/32 = 45/32 = 1.4$.

4. Scan Cell Transition Count Calculation

The paper in [6] has developed an expression, called weighted transition count (WTC), to compute in the scan chain the number of transitions caused by a test vector being scanned in. The expression is based on the observation that the number of scan cell transitions caused by a transition in a test vector being scanned in depends on its position in the test vector. The number of weighted transitions is given by:

Weighted_Transitions = $\sum$ (Size_of_Scan_Chain – Position_of_Transition) (10)

It is shown that the weighted transition count is very well correlated with the real power dissipation and hence the power dissipated when applying two vectors can be compared by counting the number of weighted transitions in the vector. However, the WTC with respect to a test vector can also be computed by counting the transitions at each scan cell. For example, consider a scan chain with five scan cells (d_1, d_2, d_3, d_4, d_5) and the test vector $b_5b_4b_3b_2b_1 = 00101$ being scanned in. The test vector has 3 transitions, T_1 between b_1 and b_2, T_2 between b_2 and b_3, and T_3 between b_3 and b_4. By Eq. (1), WTC can be computed by summing up the weight of each transition in the test vector; that is, WTC = (5-1) + (5-2) + (5-3) = 9. The same result can also be obtained by summing up the number of transitions at each scan cell in the scan chain. In the following, we compute the number of transitions at each scan cell. Note that not all the bits in a test vector will pass through every scan cell. For instance, only subsequences b_2b_1 and $b_4b_3b_2b_1$ will pass through the scan cells d_4 and d_2, respectively. To put it more formally, given a scan chain with m scan cells, the subsequence of a test vector to pass through the scan cell d_i can be expressed as $b_{m-i+1}b_{m-i}...b_1$ where $1 \leq i \leq m$. Therefore, the number of transitions at a scan cell can thus be calculated by counting the number of transitions in the subsequence that passes through it. Continuing above example, for the first scan cell d_1 there are 3 transitions in $b_5b_4b_3b_2b_1$; hence the transition count for scan cell d_1 is 3. Similarly, there are 3, 2, 1, and 0 transitions in the subsequences $b_4b_3b_2b_1$, $b_3b_2b_1$, b_2b_1, and b_1, respectively. Hence the transition counts for scan cells d_2, d_3, d_4, and d_5 are 3, 2, 1, and 0, respectively. The WTC can thus be calculated by summing up the transition count of each scan cell in the scan chain, i.e. 3 + 3 + 2 + 1 + 0 = 9. This value is the same as the value obtained by Eq. (1). What we have explained above reveals that the number of transition count at a scan cell depends not only on the number of transitions in the test vector but also on its position in the scan chain. The number of transition count caused by scanning in test vector V_t at the scan cell d_i is given by:

$$TC_{scanin}(d_i) = \sum_{j=1}^{m-i}(V_t(j) \oplus V_t(j+1)) \qquad (11)$$

where m is the length of the scan chain and $V_t(j)$ represents the jth bit of the test vector V_t. Similar reasoning can be applied to scan out test responses. The number of transition count caused by scanning out test response V_r at the scan cell d_i is given by:

$$TC_{scanout}(d_i) = \sum_{j=1}^{i-1}(V_r(j) \oplus V_r(j+1)) \qquad (12)$$

The WTC value corresponding to V_k scan-in or scan-out can be obtained by summing up transition counts of each scan cell in the scan chain and is given by:

$$WTC(V_k) = \sum_{i=1}^{m}TC(d_i) \qquad (13)$$

5. Proposed approach

Test vector ordering is equivalent to a permutation of a given set of test vectors. All the test vector ordering techniques in [1-5] are based on the construction of a transition graph with a weight associated with each edge. The weight represents the power consumed after consecutive test vectors are applied. The problem is then amounted to finding a Hamiltonian path of minimum cost in the transition graph. The main difference between these approaches is the representation of the weight. In the following we present a method to calculate the weight for each pair of test vectors. Since the power consumed during scan testing can be divided into two major parts: power during shift cycle and power during application cycle, the calculation of the weight can also be divided in two parts: weight for state input part and weight for primary input part. First, we consider the calculation of weight for the state input part. Suppose that each test vector for the CUT has $n + m$ bits where the first n bits form the primary input part and the last m bits form the state input part. When applying a test vector pair (V_{k-1}, V_k) to the CUT, the state output of V_{k-1} is scanned-out simultaneously with the scanning-in of the state input of test vector V_k. And the transition count of a scan cell d_i, $TC(d_i)$, in the scan chain after applying the test vector pair (V_{k-1}, V_k) can be expressed as:

$$TC(d_i) = \sum_{j=m+1}^{m+n-i}(R_{k-1}(j) \oplus R_{k-1}(j+1)) + \sum_{j=m+1}^{i-m-1}(V_k(j) \oplus V_k(j+1)) \qquad (14)$$

where $R_{k-1}(j)$ represents the jth bit of the state output after applying test vector V_{k-1} and $V_k(j)$ represents the jth bit of the state input of test vector V_k. As described previously, an impact function is used to measure the impact of a circuit input on the switching activity in the internal circuit. The circuit input with higher impact function value has higher probability to cause more transitions in the CUT. Hence, the weight which takes into account dependencies between internal nodes and the transitions on the state input sp_i can thus be defined as $TC(d_i) \times Isp_i$. Therefore, the weight for the state input part on each edge (V_{k-1}, V_k) representing the cost in terms of transition density of the application of the test vector pair (V_{k-1}, V_k) to the state forward cone can be defined as:

$$Weight\_sip(V_{k-1}, V_k) = \sum_{i=m+1}^{n+m}TC(d_i) \times Isp_i \qquad (15)$$

where Isp_i is the impact function value of state input sp_i and $TC(d_i)$ is the transition count of scan cell d_i after applying the test vector pair (V_{k-1}, V_k).

Secondly, we consider the calculation of weight for the primary input part. If a transition at a primary input of the CUT propagates to the internal circuit, it will subsequently cause more transitions. When applying the state input parts of two consecutive test vectors, say (V_{k-1}, V_k), to the CUT, the transition occurred at a primary input p_i, $TC(p_i)$, can be represented as:

$$TC(p_i) = (V_{k-1}(i) \oplus V_k(i)) \quad (16)$$

Hence, the weight which takes into account dependencies between internal nodes and the transitions on the primary input p_i can thus be defined as $TC(p_i) \times Ip_i$. The weight for the primary input part on each edge (V_{k-1}, V_k) representing the cost in terms of transition density of the application of the test vector pair (V_{k-1}, V_k) to the primary forward cone can be defined as:

$$Weight\_pip(V_{k-1}, V_k) = \sum_{i=1}^{n} TC(p_i) \times Isp_i \quad (17)$$

The total weight on edge (V_{k-1}, V_k) can be formally defined as:

$$\begin{aligned} Weight(V_{k-1}, V_k) &= Weight\_pip(V_{k-1}, V_k) + \\ &\quad Weight\_sip(V_{k-1}, V_k) \\ &= \sum_{i=1}^{n} TC(p_i) \times Isp_i + \sum_{j=m+1}^{n+m} TC(d_j) \times Isp_j \end{aligned}$$
$$(18)$$

Similar to the other test vector ordering approaches in [1-5], construction of a transition graph is required in this paper. Consider a set of test vectors $S = (V_1, V_2, ..., V_n)$. We construct a complete directed graph $G = (\varphi, E)$, where each vertex $V_k \in \varphi$ represents a test vector of S and each directed edge $(V_i, V_j) \in E$ represents a pair of test vectors. The weight on each edge (V_i, V_j) of G is calculated using Eq. (18). Then the test vector ordering problem is amounted to finding a Hamiltonian path of minimum cost in the complete graph, which is known to be an NP-hard problem. A rich literature exists on the algorithms for solving this problem. We develop a program based on the algorithm in [8] to determine the order of the test vectors that minimizes the power consumed during test.

6. Experimental results

To validate the proposed approaches, we have carried out experiments on full scan versions of the ISCAS 89 benchmark circuits. Table 2 shows basic characteristics of these circuits. They are gate counts, number of Flip-Flops, primary inputs and primary outputs, respectively. The procedure to determine the order of the test vectors was implemented on a 1.5 GHz Pentium IV PC with 512 MB RAM running Linux and using GNU CC version 2.19. We use average transition count (ATC) as quantitative measure for power consumption. The ATC is defined as the total number of transition counts divided by the total number of clock cycles. To evaluate the effectiveness of the proposed vector ordering approach, we compare the ATC and the improvement of transition reduction with that of the simulation based approach. Simulation based approach is founded on exhaustive counting of the transitions at all circuit lines for every test pair. The calculation based test vector ordering approach in paper [5] are also implemented and are compared with the proposed approach. The first column gives the circuit name. In the second column, the first sub-column *Original* shows the ATC for the circuits without using any vector ordering techniques. The second, third, and fourth sub-columns (*Our*, *Paper*[5], and *Simu.*) show the ATC by using the proposed approach, the approach in paper [5], and the simulation based approach, respectively. The next column shows the improved percentages of ATC for the proposed approach, the approach in paper [5], and the simulation based approach, respectively. Compared with the approach without using any vector reordering techniques, only 8.18% of power reduction is improved for the proposed approach. This is because lots of the circuits in Table 3 are controlled mostly by scan chains rather than primary inputs. However, the test vector ordering approach is independent to a lot of approaches such as circuit modification, scan chain partitioning, and scan cell ordering, so it can be utilized together with these approaches to further reduce test power. The last column compares the proposed approach with the simulated approach in terms of *improvement efficiency* which is defined as the ratio between the improved percentages obtained by the proposed approach (or the approach in [5]) and by the simulation based approach. From Table 3, we find that the percentage improvement achieved by the proposed approach is close to the simulation based approach and the improvement efficiency can achieve 91.55% on average and always has better results than the approach

Table 2. Characteristics of ISCAS 89 benchmark circuits

Circuit	Gates	FF's	PI's	PO's
s298	119	14	3	6
s344	160	15	9	11
s420	218	16	19	2
s510	211	6	19	7
s641	379	19	35	24
s713	393	19	35	23
s1423	657	74	17	5
s5378	2779	179	35	49
s9234	5597	211	36	39
s13207	7951	638	62	152
s15850	9772	534	77	150
s35932	16065	1728	35	320
s38417	22179	1426	38	304

Table 3. Results of power reduction

Circuit	Average Transition Count				Improvement (%)			Efficiency (%)		CPU Time (Sec.)
	Original (A)	Ours (B)	Paper[5] (C)	Simu. (D)	Ours $\frac{(A)-(B)}{(A)} \times 100$	Paper[5] $\frac{(A)-(C)}{(A)} \times 100$	Simu. $\frac{(A)-(D)}{(A)} \times 100$	Ours	Paper [5]	
s298	140.61		132.732	130.06	7.37	5.60	7.5	98.27	74.70	0.14
s344	293.38		273.207	265.10	8.84	6.88	9.64	91.70	71.33	0.31
s420	307.3		283.681	270.49	9.86	7.69	11.98	82.30	64.16	0.83
s510	314.42		269.239	258.61	16.6	14.37	17.75	93.52	80.96	0.34
s641	426.65		383.615	374.39	10.73	10.09	12.25	87.59	82.34	1.17
s713	409.48	362.27	366.042	358.01	11.53	10.61	12.57	91.73	84.39	1.21
s1423	1019.25	937.00	953.875	933.84	8.07	6.41	8.38	96.30	76.54	10.75
s5378	3883.39	3580.49	3656.18	3557.96	7.8	5.85	8.38	93.08	69.82	38.54
s9234	6758.39	6335.99	6468.31	6315.04	6.25	4.29	6.56	95.27	65.43	76.42
s13207	12579.98	130.25	12109.3	11805.05	5.59	3.74	6.16	90.75	60.74	109.24
s15850	11079.91	267.45	10681.6	10427.30	5.21	3.59	5.89	88.46	61.03	98.87
s35932	22784.8	277.00	22142.2	21613.66	4.83	2.82	5.14	93.97	54.87	35.29
s38417	33023.1	262.23	32308.7	31652.64	3.62	2.16	4.15	87.23	52.13	1473.2
Avg.	7155.44	380.87	6925.28	6766.32	8.18	6.47	8.95	91.55	69.11	142.03

in [5] in all cases. In simulation based approach the calculation of ATC for every possible pair of test vectors requires logic simulation. When the test set is large the required computation time is prohibitively long, although it has better power reduction rate than that of the calculation based approach.

7. Conclusions

We have presented a calculation based test vector ordering approach to reduce power dissipation for full scan sequential circuits. The proposed approach exploits the dependencies between internal circuits and the transitions at circuit inputs. Two functions are developed to compute the transition weight of a scan cell and to measure the impact of transitions at a circuit input on switching activity in the internal circuit, respectively. Experimental results for benchmark circuits show that improvement efficiency can achieve 91.55% on average.

8. References

[1] V. Dabholkar, S. Chakravarty, I. Pomeranz, and S. Reddy, Techniques for minimizing power dissipation in scan and combinational circuits during test application, *IEEE Trans. Computer-Aided Design*, Dec.1998, vol. 17, pp. 1325-1333,.

[2] Z. Luo, X. Li, H. Li, S. Yang, and Y. Min, Test Power Optimization Techniques for CMOS Circuits, 11th *Asia Symposium*, 2002, pp. 332-337.

[3] X. Kavousianos, D. Bakalis, M. Bellos, and D. Nikolos, An Efficient Test Vector Ordering Method for Low Power Testing, *Proceedings IEEE Computer Society Annual Symposium on VLSI*, Feb. 2004, pp. 285-288.

[4] P. Girard, C. Landrault, S. Pravossoudovitch, and D. Severac, Reducing power consumption during test application by test vector ordering, *Proc. IEEE int'l Symp. Circuits and Systems*, part II, 1998, pp. 296-299.

[5] P. Girard, L. Guiller, C. Landrault, and S. Pravossoudovitch, A Test Vector Ordering Technique for Switching Activity Reduction during Test Operation, *Proceedings Ninth Great Lakes Symposium on VLSI*, Mar. 1999, pp. 24-27.

[6] R. Sankaralingam, R. R. Oruganti, and N. A. Touba, Static compaction techniques to control scan vector power dissipation, *VLSI Test Symposium*, 2000, pp. 35-40.

[7] S. Wang and S. K. Gupta, DS-LFSR: a BIST TPG for low switching activity, *IEEE Trans. Computer-Aided Design*, July 2002, vol. 21, pp. 842-851.

[8] M. N. Swamy and K. Thulasiraman, *Graphs, Networks, and Algorithms*. New York: Wiley, 1981.

Formal Methods
MTV 2007

Mechanized Certification of Secure Hardware Designs

Sandip Ray
Department of Computer Sciences
University of Texas at Austin
Austin, TX 78712. USA.
sandip@cs.utexas.edu

Warren A. Hunt, Jr.
Department of Computer Sciences
University of Texas at Austin
Austin, TX 78712. USA.
hunt@cs.utexas.edu

Abstract— **We develop a framework for mechanized certification of secure hardware systems built out of commercial off-the-shelf (COTS) components purchased from untrusted vendors. Certification requires a guarantee that the fabricated system satisfies the requisite safety and security properties. Our framework facilitates this by (1) providing an unambiguous description of the requirements specification in a formal, computational logic, (2) a formalized hardware description language (HDL) to describe the implementation, and (3) mechanical tools and techniques for providing a certification of correctness and security. We illustrate the use of the framework in certifying the correctness and security properties of the netlist implementation of a voting machine using the ACL2 theorem prover.**

I. Introduction

Society today is dependent on computing systems built out of commercial off-the-shelf (COTS) components for critical applications. It is therefore crucial that such systems perform in a secure and reliable manner. Consequently there has recently been significant interest, both in the government and in the industry, in developing mechanized tools for certification of security-critical systems built out of COTS components.

Certification of a COTS system requires interaction between designers, consumers, and regulatory evaluators to guarantee that the fabricated system satisfies the requisite safety and security properties. To achieve this, one must have (1) an unambiguous specification of the system requirements, (2) a formal semantics of the language in which the system is implemented, and (3) mechanized mathematical tools to assist the analyst in the certification process.

We are developing a framework, based on machine-assisted formal reasoning, to facilitate mechanized certification of secure hardware designs. Our motivation is to provide the following key ingredients:

- A formal mathematical logic as a basis for requirements specification.
- A formalized Hardware Description Language (HDL) as a means for describing hardware implementation.
- Use of scalar and symbolic simulation, and automated formal analysis, on the same formal artifact.

In this paper we describe different facets of our framework, using, as a simple illustrative example, its application in the mechanized certification of the netlist implementation of a voting machine. The process illustrates the nature of mechanized infrastructural support necessary for the certification of security-critical hardware designs.

The formal foundational basis for our framework is provided by the ACL2 system [12]. ACL2 is a general-purpose theorem prover based on an applicative subset of Common Lisp. ACL2 has been successfully used in the formal analysis of a slew of computing systems, ranging from pipelined microprocessors to JVM byte codes [18], [20], [22], [16]. In our framework we make critical use of the mechanical reasoning engine of ACL2, and in particular its support for efficient function execution which facilitates validation of the formal models by simulation. However, the key ideas are independent of the nuances of a specific theorem proving system.

The rest of the paper is organized as follows. In Section II, we provide a brief overview of the ACL2 logic and theorem prover. Section III deals with the development of specification of a hardware device. In Section IV, we discuss a formalized hardware description language (HDL) called **DE**, and show how it can be used to succinctly and unambiguously describe hardware implementations. In Section V, we describe the key analysis steps in a certification, pointing out the respective roles of simulation and formal reasoning; we also discuss how our approach affords effective computation of information flow properties of the system. We conclude the paper in Section VI. A modicum of familiarity with Lisp is assumed in the presentation. However, no previous familiarity with ACL2 is required; the relevant aspects of ACL2 are mentioned in Section II.

II. Overview of ACL2

In this section, we briefly describe the ACL2 logic. This provides a formal notational and reasoning framework to be used in the rest of the paper. We refer the reader interested in a thorough understanding of ACL2 to the ACL2 Home Page (http://www.cs.utexas.edu/users/moore/acl2), which contains an extensive hypertext documentation and pointers to several published books and papers.

ACL2 is a first-order logic of recursive functions. The inference rules constitute propositional calculus with equality and instantiation, and well-founded induction up to ϵ_0. The language is an applicative subset of Common Lisp; instead of writing $f(a)$ as the application of function f to argument a, one writes (f a). Terms are used instead of formulas. For example, the following term represents a basic fact about lists in the ACL2 syntax.

```
(implies (natp i)
         (equal (nth i (update-nth i v l))
                v))
```

The syntax is quantifier-free; formulas may be thought of as universally quantified over all free variables. The term above specifies the statement: "For all i, v and l, if i is a natural number, then the i-th element of the list obtained by updating the i-th element of l by v is v."

ACL2 provides axioms to reason about Lisp functions. For example, the following axiom specifies that the function car applied to the cons of two arguments, returns the first argument of cons.

Axiom:
(equal (car (cons x y)) x)

The Lisp axioms of ACL2 together constitute the ACL2 *Ground Zero Theory* ($\mathcal{GZ}$ for short). $\mathcal{GZ}$ characterizes about 170 functions described in the Common Lisp Reference Manual [23], which are (i) free from side effects, (ii) independent of the state or other implicit parameters or data types other than those supported by ACL2, and (iii) unambiguously specified on their intended domains in a host-independent manner. The return values predicted by the axioms agree with those specified in the Common Lisp Manual for arguments in the intended domains.

Theorems can be proved for axiomatically defined functions in the ACL2 system. Theorems are proved by the defthm command. For example, the command:

```
(defthm car-cons-for-2
  (equal (car (cons x 2)) x))
```

directs the theorem prover to prove that for every x, the output of the function car applied to the cons of x and the constant 2, returns x.

ACL2 provides *extension principles* that allow the user to introduce new function symbols and axioms about them. The extension principles constitute (i) the *definitional principle* to introduce total functions, (ii) the *encapsulation principle* to introduce constrained functions, and (iii) the *defchoose principle* to introduce Skolem functions. We briefly sketch these principles here.[1] Kaufmann and Moore [14] present a detailed description of these principles along with their soundness arguments. Any ACL2 theory is an extension of $\mathcal{GZ}$ through applications of the extension principles.

Definitional Principle:: The *definitional principle* allows the user to define new total functions. For example, the following form defines the factorial function fact in ACL2.

---

[1] ACL2 has another extension principle, namely the *defaxiom principle*, which permits specifying any formula in the current theory as an axiom. The use of this principle is discouraged since can lead to inconsistent theories. We ignore defaxiom principles in our framework.

```
(defun fact (n)
  (if (zp n)
      1
      (* n (fact (- n 1)))))
```

The effect is to extend the logic by the following *definitional axiom*:

Definitional Axiom:
(fact n)
=
(if (zp n) 1 (* n (fact (- n 1))))

Here (zp n) returns nil if n is a positive natural number, and otherwise T. To ensure consistency, ACL2 must prove that the recursion terminates [5]. In particular, one must exhibit a "measure" m that maps the set of arguments in the function to some set W, where $\langle W, \prec \rangle$ forms a well-founded structure. The proof obligation, then, is to show that on every recursive call, this measure "decreases" according to relation $\prec$. ACL2 axiomatizes a specific well-founded structure, namely the set of ordinals below ϵ_0: membership in this set is recognized by an axiomatically defined predicate o-p, and a binary relation o< is axiomatized in the logic as an irreflexive partial order in the set.

Encapsulation Principle:: The *encapsulation principle* allows the extension of the ACL2 logic with partially defined constrained functions. For example, the command below introduces a function symbol foo with the constraint that (foo n) is a natural number.

```
(encapsulate
  (((foo *) => *))
  (local (defun foo (n) 1))
  (defthm foo-returns-natural
    (natp (foo n))))
```

Consistency is ensured by showing that some (total) function exists satisfying the alleged constraints. In this case, the constant function that always returns 1 serves as such "witness". The effect is to extend the logic by the following *encapsulation axiom* corresponding to the constraints. Notice that the axiom does not specify the value of the function for every input.

Encapsulation Axiom:
(natp (foo n))

For a constrained function f the only axioms known are the constraints. Therefore, any theorem proved about f is also valid for a function f' that also satisfies the constraints. More precisely, call the conjunction of the constraints on f the formula ϕ. For any formula ψ let $\hat{\psi}$ be the formula obtained by replacing the function symbol f by the function symbol f'. Then, a derived rule of inference, *functional instantiation* specifies that for any theorem θ one can derive the theorem $\hat{\theta}$ provided one can prove $\hat{\phi}$ as a theorem. In the example, since the constant 10 satisfies the constraint for foo, if (bar (foo n)) is provable for some function bar, functional instantiation can be used to prove (bar 10).

Defchoose Principle:: The *defchoose principle* allows introduction of Skolem functions in ACL2. To understand this principle, assume that a function symbol P of two arguments has been introduced in the ACL2 logic. Then the form:

```
(defchoose exists-y-witness y (x)
  (P x y))
```

extends the logic by the following axiom:

Defchoose Axiom:
```
(implies (P x y)
         (P x (exists-y-witness x)))
```

The axiom states that *if* there exists some y such that (P x y) holds, then (exists-y-witness x) returns such a y. Nothing is claimed about the return value of (exists-y-witness x) if there exists no such y. This provides the power of first-order quantification in the logic. For example, we can define a function exists-y such that (exists-y x) is true if and only if there exists some y satisfying (P x y). Notice that the theorem exists-y-suff below is an easy consequence of the defchoose and definitional principles.

```
(defun exists-y (x)
  (P x (exists-y-witness x)))

(defthm exists-y-suff
  (implies (P x y) (exists-y x)))
```

ACL2 provides a construct defun-sk that makes use of the defchoose principle to introduce explicit quantification. For example, the form:

```
(defun-sk exists-y (x)
  (exists y (P x y)))
```

is merely an abbreviation for the following forms:

```
(defchoose exists-y-witness y (x)
  (P x y))

(defun exists-y (x)
  (P x (exists-y-witness x)))

(defthm exists-y-suff
  (implies (P x y)
           (exists-y x)))
```

Thus (exists-y x) can be thought of specifying as the first-order formula: $(\exists y : (P\ x\ y))$. Further, defun-sk supports universal quantification forall by exploiting the duality between existential and universal quantification.

A. Executability in ACL2

ACL2 provides strong support for executing functions introduced through the definitional principle. To support this, the invocation of a definitional principle in ACL2 entails performing several operations, in addition to introducing the definitional axiom. In particular, a new function symbol is defined (and generally compiled) in the host Common Lisp. For example, the above defun for fact is executed directly in Common Lisp. We refer to this definition as the *executable counterpart* of fact. Because Common Lisp is a model of the ACL2 axioms, ACL2 may exploit the Common Lisp counterpart and the host Lisp execution engine as follows: when a ground application of the defined symbol arises during the course of a proof or when the user submits a form to ACL2, its value under the axioms may be computed with the Common Lisp counterpart in the host Lisp. For example, should (fact 15) arise in a proof, ACL2 can use the Common Lisp counterpart of fact to compute 1307674368000 in lieu of deriving that value by repeated reductions using instantiation of the definitional axioms.

The story above is made somewhat more subtle by the fact that the Common Lisp functions are partial, while ACL2 functions are total. For instance, in Common Lisp, (car 7) is undefined while in the ACL2 logic the value is provably NIL. ACL2 provides a mechanism, called *guards* [13] to enable the use of the Common Lisp counterpart only on ground terms where the arguments for each function f are in the intended domain of application of f. ACL2 contains contains several other constructs to support efficient executability, such as (1) single-threaded objects [6], and (2) mbe [9]. Single-threaded objects enable destructive updates to certain data structures in an applicative context. Mbe (or must-be-equal) allows the user to attach different function bodies to the same function definition; one body is used for logical reasoning and the other for executability, and the user proves the logical equality of their return values using the theorem prover.

III. SPECIFICATION

The first crucial step in the development of a certifiable design is the definition of its *specification*. Unfortunately, in current practice, little attention is given to making the specification formal or unambiguous: system requirements are typically described with charts and diagrams, together with ambiguous English. In this section we will discuss an informal specification of the requirements of a voting machine, point out the the inadequacy of such a description as a basis of certification, and then show how to refine such descriptions to a formal specification that can be mechanically analyzed.

An informal description of the voting machine is as follows. The machine has a counter for each candidate. At any instant it has status :ready, :locked, or :frozen, and responds to the following user actions:

- At the :ready state, the voter performs a :vote action to tentatively select a candidate. The system records the vote, but does not change state.
- The voter can change her mind by performing :reset. This clears the tentative selection above.
- Once the :commit action is selected, the system records the vote and transits to the state :locked.
- The :unlock action is performed by a polling official after a vote has been cast and the voter left. The system changes state from :locked to :ready.

- The `:freeze` action is performed when polling is completed. The machine then provides a tally of votes.

The above sounds like a reasonable description of the specification of a voting machine. Nevertheless, it is not hard to find omissions. For instance, what should happen if `:unlock` is performed when the machine is `:ready`? We tacitly assumed that `:unlock` occurs only in the `:locked` state. This requires that (i) the voter does not leave without casting a vote, or (ii) if she does then the polling official does not unlock the machine.

To avoid omissions, we define specifications operationally with (i) the initial state of the machine, and (ii) a state transition function (`spec` s i) which defines the next state for each state s and input i. Attempting to formalize the above description immediately detects our omission: since ACL2 functions are total, `spec` must define the next state when `:unlock` is performed when the machine is `:ready`. We therefore refine the specification by stipulating that in this case the machine clears the tentative votes of the undecided voter. We further stipulate that if it encounters an "unexpected" input at any state, no change of state occurs. A fragment of our formalized `spec` function with these stipulations is shown in Fig. 1.

A specification requires consideration of the possible system behaviors and involves several design choices: instead of rejecting an unexpected input, an alternative could be for the machine to transit to an error state. Note that we can use encapsulation in ACL2 so that `spec` is defined only for expected inputs. While this approach is sometimes convenient, we prefer executable definitions whenever possible, since it facilitates validation of the specification via simulation.

The specification above is defined as a state machine rather than by formulas representing properties of the implementation. In addition to simulation, this affords intuitive specifications in practice. Most implementations are elaborations of simpler protocols to achieve execution efficiency, match a given architecture, etc. The simpler protocol then succinctly captures the behaviors of the elaboration. On the other hand, most modern systems are *reactive* and their properties are naturally described in a temporal logic. Defining such formulas thus requires a semantic embedding of temporal logic, which is cumbersome because of the first-order nature of ACL2 [19]. However, to use operational specifications, we must additionally formalize a notion of correspondence between the state machines. We address this in Section V.

IV. IMPLEMENTATION

Having described the specification of our voting machine, we turn now to our approach to describing the implementation in a way that affords mechanized certification.

In order to certify that an implementation satisfies a specification, the implementation must be represented in a language with formal and unambiguous semantics. However, in practice, hardware designs are typically implemented in some commercial Hardware Description Language (HDL) such as VHDL [4] and Verilog [24]. These HDLs need to satisfy several disparate

```
(defun s-init ()
  (>_ :status :ready
      ...))

(defun spec (s i)
  (let ((satus (status s))
        (c0 (candidate0 s))
        (c1 (candidate1 s))
        (opcode (opcode i)))
  (case opcode
    (:vote
      (case status
        (:ready
          (case (candidate i)
            (0 (>s :tvote0 1
                   :tvote1 0))
             ...))
        ...
        (t s)
    (:commit
      (case status
        (:ready
          (>s :candidate0
              (+ c0 (tvote0 s))
              :candidate1
              (+ c1 (tvote1 s))
              :status :locked))
          ...
          (t s)))
    (:unlock
      (case status
        (:ready (>s :tvote0 0
                    :tvote1 0))
         ...
         (t s)))
    (:freeze
       (>s :status :frozen
           :tally ...))
    (t s))))
```

Fig. 1. Fragment of a Voting Machine Specification. Here we use the ACL2 records book [15] to update and access machine components; (`status s`), (`opcode i`), etc., are accessors, `>s` is a macro for updating fields of record `s`, and `>_` updates the empty record.

goals other than formal verification, namely ease of use, simulation speed, etc. As a result, most commercial HDLs are large, unwieldy, and in parts poorly specified [21]. Therefore, formal analysis of a hardware design written in a commercial HDL has been traditionally restricted to some alternative encoding of the underlying algorithm written (typically by a human) in some formal language. The utility of such a verification then rests upon the assumption that the encoding faithfully reflects the actual implementation.

Our solution to this problem is the development of the **DE** language [11].[2] **DE** is a hierarchical, occurrence-oriented HDL with a formal semantics defined by a deep embedding in the logic of ACL2.

Figure 2 shows a fragment of the netlist representation of our voting machine. Note that the netlist is represented as a constant (declared by the `defconst` construct) in the ACL2 logic. The name of the constant is `*vnlst*`. The netlist has five modules `vote`, `status`, `cmtvote`, `4-bit-ctr`, and `1-bit-ctr`. A module has input and output wires, state holding elements, and a set of occurrences; module `cmtvote` has three inputs (`candidate`, `commit`, and `reset-`), eight outputs (`out00`, `out01`, etc.), two state elements (`vote0` and `vote1`), and five occurrences (`vote0`, `vote1`, `g0`, `g1`, and `g2`). Some modules like `and`, `not`, etc., are *primitive*. In other modules, *occurrences* describe connections by instantiating other modules: in `cmtvote`, `g2` represents connection of (input) wire `candidate` and (internal) wire `ncandidate` by instantiating the `not` module. The top module `vote` instantiates two modules `status` and `cmtvote`; `vote` has input bits `op0`, `op1`, and `op2` encoding user actions, and a `candidate` input. The `status` module updates the status values **:ready**, **:locked**, etc., encoded in two state bits. Module `cmtvote` updates vote counts. Counting is done using 4-bit counters for demonstration purposes.

We now discuss the semantics of the **DE** language. The semantics is provided by defining a formal language interpreter in the ACL2 logic. The interpreter functions `se` and `de` are shown in Fig. 3. Function `se` returns the outputs of a module `fn` of a netlist `n` as a function of its inputs and state elements, and `de` returns the next state. Here `primp` determines if `fn` is a primitive module; `se-primp-apply` and `de-primp-apply` are primitive module evaluators; `se` crawls over the module structure recursively evaluating each signal occurrence and finally filtering the outputs; `de` performs a second pass to evaluate the next states. Note that unlike commercial HDLs, **DE** has a compact semantics: the above definitions together with the primitive evaluators constitute the *entire* language definition. The regularity and economy of **DE** makes it suitable for mechanically analyzable hardware implementations.

V. ANALYSIS AND DISCUSSIONS

The key analysis step involves showing that the executions of the netlist satisfy the specification. Since the specification itself is a state machine, formalizing this requires a notion of correspondence between two state machine executions.

We formalize correspondence with functions `rep`, `good`, `pick`, and `inv` so that the formulas in Fig. 4 are theorems. The theorems imply that every `good` execution of the implementation is matched by the specification and essentially formalize the notion of *trace containment* [2] in ACL2, where

---
[2]The **DE** language is the successor of the DUAL-EVAL HDL [7], and is an evolving project [10], [11]. The version of **DE** used in the analysis described here is called **DE4**. Recently, Boyer and Hunt have developed a version of the language called **E**, with a number of sophisticated analysis capabilities.

```
(defconst *vnlst*
 '((vote
    (op0 op1 op2 candidate)
    (sout0 sout1
     out00 out01 out02 out03
     out10 out11 out12 out13)
    (votes stat)
    ((stat (sout0 sout1)
           status
           (op0 op1 op2))
     (votes (out00 out01 out02
             out03 out10
             out11 out12 out13)
            cmtvote
            (candidate commit reset))
     ...))
   (status
    (op0 op1 op2)
    (sout0 sout1)
    (s0 s1)
    (...))
   (cmtvote
    (candidate commit reset-)
    (out00 out01 out02 out03
     out10 out11 out12 out13)
    (vote0 vote1)
    ((vote0 (out00 out01 out02 out03)
            4-bit-ctr
            (commit0 reset-))
     (vote1 (out10 out11 out12 out13)
            4-bit-ctr
            (commit1 reset-))
     (g2 (ncandidate)
         not
         (candidate))
     (g0 (commit0)
         and
         (commit ncandidate))
     (g1 (commit1)
         and
         (commit candidate))))
   (4-bit-ctr
    (incr reset-)
    (out0 out1 out2 out3)
    (h0 h1 h2 h3)
    ((h0 (out0 carry0)
         1-bit-ctr
         (incr reset-))
     (h1 (out1 carry1)
         1-bit-ctr
         (carry0 reset-))
     (h2 (out2 carry2) ...)
     (h3 ...)))
   (1-bit-ctr ...)))
```

Fig. 2. Fragment of the netlist representation of a voting machine.

```
(mutual-recursion
 (defun se (fn ins sts n)
  (if (primp fn)
      (se-primp-apply fn ins sts)
    (let ((m (assoc-eq fn n)))
     (if (atom m) nil
       (assoc-eq-values
        (md-outs m)
        (se-occ (md-occs m)
                (pairlis$ md-ins ins)
                (pairlis$ md-sts sts)
                (delete-eq-module
                 fn n))))))))
 (defun se-occ (occs w-alst s-alst n)
  (if (endp occs) w-alst
    (let* ((occ (car occs))
           (ins (assoc-eq-values
                 (occ-ins occ)
                 w-alst))
           (sts (assoc-eq-value
                 (occ-name occ)
                 s-alst)))
      (se-occ (cdr occs)
              (append
               (pairlis$ (occ-outs occ)
                         (se (occ-fn occ)
                             ins sts n))
               w-alst)
              sts n)))))

(mutual-recursion
 (defun de (fn ins sts n)
  (if (primp fn)
      (de-primp-apply fn ins sts)
    (let ((m (assoc-eq fn netlist))
          (n-n (delete-eq-module fn n)))
     (if (atom m) nil
       (assoc-eq-values md-sts
        (de-occ (md-occs m)
         (se-occ
          (md-occs m)
          (pairlis$ (md-ins m) ins)
          (pairlis$ (md-sts m) sts)
          n-n)
         (pairlis$ (md-sts m) sts)
         n-n))))))
 (defun de-occ (occs w-alst s-alst n)
  (if (endp occs) w-alst
    (let* ((occ (car occs))
           (ins (assoc-eq-values
                 (occ-ins occ)
                 w-alst))
           (sts (assoc-eq-value
                 (occ-name occ)
                 s-alst)))
      (de-occ (cdr occs)
              (acons (occ-name occ)
                     (de (occ-fn occ)
                         ins sts n)
                     w-alst)
              s-alst n)))))
```

Fig. 3. Definition of Semantics for **DE** Language

```
(defthm rep-matches
 (and
  (equal (rep *init*)
         (s-init)))
  (implies
    (and (inv s)
         (good s i))
    (equal
     (rep (de 'vote s i *vnlst*))
     (spec (rep s) (pick i)))))

(defthm inv-invariant
 (and
  (inv *init*)
  (implies
    (and (inv s)
         (good s i))
    (inv (de 'vote s i *vnlst*)))))
```

Fig. 4. Theorems showing that the netlist implementation of the voting machines is a refinement of spec. Here *init* is the valuation of the state elements at the initial state, rep maps the design states to specification states, pick is the input mapping, inv is an invariant on the design, and (good s i) checks if i is a valid design input at state s.

containment is restricted to good traces.[3]

Note that the specification needs to match the implementation *only* for good transitions. Contrast this with our approach of defining spec as a total function. While we could similarly insist that the specification must match *each* implementation step, this often complicates definitions. For instance, spec uses unbounded additions above while *vnlst* uses 4-bit counters. Modifying spec to use bounded arithmetic would complicate its definition, and furthermore, the definition would no longer be applicable if we re-design the netlist with (say) 64-bit counters. We prefer generic specifications and use good to impose input constraints.

Proving the above theorems is a two-step process. The first step is what is referred to as *semantic simplification*. In this step, we define functions in ACL2 (called *semantic functions*) that mimic the workings of each module, and prove theorems relating the se and de expressions with these functions. Fig. 5 shows the theorems for the module 4-bit-ctr.

The theorems relating se and de expressions are proven hierarchically. Since 4-bit-ctr instantiates 1-bit-ctr, we first prove analogous theorems for the latter; the theorems shown in Fig. 5 are then proven by symbolic expansion of se and de functions and applying the 1-bit-ctr theorem for the corresponding occurrence. The process can be automated with Lisp macros [11].

In the second step we define rep, good, pick, and inv. The first three definitions are typically easy; for instance, rep maps the bit configurations of state elements stat and votes to keyword-based status values and numerical vote

---
[3]It is sometimes more convenient to use trace containment under *stuttering* to relate to machines at different abstractions [17]. We do not discuss stuttering in this paper.

```
(defun 4btnt (n)
  (and
    (equal (assoc-eq '4-bit-ctr n)
           '(4-bit-ctr (incr reset-)
                        ...))
    (1btnt
      (delete-eq-module '4-bit-ctr n))))

(defthm 4-bit-ctr-se-eval
  (implies
    (4btnt n)
    (equal (se '4-bit-ctr ... n)
           ...)))
(defthm 4-bit-ctr-de-eval
  (implies
    (4btnt n)
    (equal (de '4-bit-ctr ... n)
           ...)))
```

Fig. 5. Semantic simplification of `4-bit-counter` module. The "..." at the right hand side of each equality contains an ACL2 semantic function for the behavior of the module.

counts. Theorem `rep-matches` requires showing correspondence between single steps of two machines; by virtue of our having performed the first step, proving this does not involve reasoning about the **DE** semantics. A harder problem in practice is defining `inv` and proving `inv-invariant`; the theorem shows that `inv` is an inductive invariant, and allows us to assume `inv` in the proof of `rep-matches`. The standard approach is to define a predicate `suff` that is sufficient to prove `rep-matches`; we then incrementally strengthen `suff` to an inductive invariant. However, since netlists are *finite* state machines, decision procedures can be used to derive such invariance, and recent work integrating **DE** with SAT-solvers [11] provides substantial automation.

The definitions of `rep`, `good`, and `pick` are integral parts of system specification. For purposes of certifying the implementation, it is convenient to view them as "usage instructions" supplied by a seller of a component to augment the specification provided by the buyer. For instance `rep` describes how the valuations of the netlist state elements are viewed as abstract states. Certification then requires us to (i) check the validity of the theorems above (possibly with assistance from the seller), and (ii) validate that the usage instructions do indeed correspond to the environments in which the design is deployed. Note that it is possible to have buggy implementations satisfying the theorems above, under the wrong definitions of `rep`, `good`, etc. To illustrate this, consider a machine that clears the votes of all candidates when encountering the sequence **:vote**, **:reset**, **:reset**. If the predicate `good` specifies that such a sequence does not occur then it is possible to "verify" the implementation, although the implementation is obviously buggy. The problem is that the environmental assumptions used in the verification (in particular the predicate `good`) are falsified in the deployment environment. Checking such violations involves simulation of both the usage functions (in this case the definition of input constraint `good`) and the definition of `spec`.

We now consider regulatory checks. Regulatory checks are different from functional correctness, for instance requiring the guarantee of of privacy, absence of trapdoors, etc. One can (and often does) apply theorem proving to prove such properties. However, sometimes we can do automatic checks of structural or information-flow properties by computation.

An information-flow property we prove using computation is that the votes of one candidate do not affect those of the other. To check this property, we need a cone-of-influence analysis. **DE** facilitates such analysis by the following observation.[4] Note from the definitions of `se` and `de` that the core interpreter semantics is given by the primitive evaluation functions `se-primp-apply` and `de-primp-apply`; the remainder of the definitions involves recursive crawling over the netlist. Thus we can define a different interpreter by modifying the primitive evaluators. For cone-of-influence, we modify them to return a list of the *state bits necessary for evaluation* rather than the evaluation itself; the check then involves recognizing that all state bits are included in the evaluation of `votes`. The same function is used to show that there is no "hidden state" in the netlist that does not pertain to vote counts. This guarantees the absence of trapdoors.

VI. CONCLUSION

Developing COTS systems in current practice contain a number of informal components, namely requirement description as text, graphs, and charts, implementations in languages with incomplete or complicated semantics, and incomplete testing as the primary validation procedure. This does not afford a repeatable, mechanical means to verify the security and correctness of a delivered design. Recently there has been interest in a uniform formal framework to guarantee high assurance in correct and secure system executions. The Common Criteria [1] requires a uniform *lingua franca* for communication of designers, consumers, and evaluators. Rockwell Collins has used ACL2 to achieve the highest level of assurance (EAL7) provided for by the Common Criteria in the AAMP7™ processor design [8]. We have found that ACL2 is well-suited to serve as a mechanized framework for designing high-assurance systems for several reasons. The language of ACL2 is a programming language, namely Applicative Common Lisp, which facilitates implementation of different analysis tools in the same formal framework; secondly, the logic has high execution support; third, the theorem prover has been extensively used in the verification of systems at different levels of abstraction [3]. However, we believe that it is possible to port the framework to any other theorem prover that provides strong support to executability and symbolic rewriting.

---

[4]Currently, in the **E** language, this observation has been used to develop different built-in interpreters of the same module, including information-flow interpreter discussed here.

We have illustrated an approach to mechanize the different facets of mechanized certification of the implementation of a security-critical artifact in ACL2. The formal language provides a basis for unambiguous communication among the different parties. Deep embedding enables the use of different analysis tools to be applied to the same design artifact, namely a netlist, within the same formal system. Executable specifications and usage functions afford easy requirements validation via simulation. Refinements facilitate compositional proofs of correspondence via single-step theorems. Note that all these individual steps have been extensively studied by the formal methods community; our approach shows how to effectively orchestrate the steps in increasing assurance in correct executions of highly secure systems.

The definition of our framework is under development. In future work, we plan to provide more automation in specification design. We are also planning to apply the paradigm to design integrity checks on binary code developed for practical machine architectures.

VII. Acknowledgements

This material is based upon work supported in part by DARPA and the National Science Foundation under Grant No. CNS-0429591. We thank Matt Kaufmann for several illuminating comments and suggestions in course of this research.

References

[1] Common Criteria for Information Technology Security Evaluation. See URL: http://csrc.nist.gov/cc/CC-v2.1.html.
[2] M. Abadi and L. Lamport. The Existence of Refinement Mappings. *Theoretical Computer Science*, 82(2):253–284, May 1991.
[3] W. R. Bevier, W. A. Hunt, Jr., J S. Moore, and W. D. Young. An Approach to System Verification. *Journal of Automated Reasoning*, 5(4):409–530, December 1989.
[4] J. Bhasker. *A VHDL Primer*. Prentice-Hall, 1992.
[5] R. S. Boyer and J S. Moore. *A Computational Logic*. Academic Press, New York, NY, 1979.
[6] R. S. Boyer and J S. Moore. Single-threaded Objects in ACL2. In S. Krishnamurthy and C. R. Ramakrishnan, editors, *Practical Aspects of Declarative Languages (PADL)*, volume 2257 of *LNCS*, pages 9–27. Springer-Verlag, 2002.
[7] B. Brock and W. A. Hunt, Jr. The Dual-Eval Hardware Description Language and Its Use in the Formal Specification and Verification of the FM9001 Microprocessor. *Formal Methods in Systems Design*, 11(1):71–104, 1997.
[8] D. Greve, R. Richards, and M. Wilding. A Summary of Intrinsic Partitioning Verification. In M. Kaufmann and J S. Moore, editors, *5th International Workshop on the ACL2 Theorem Prover and Its Applications (ACL2 2004)*, Austin, TX, November 2004.
[9] D. A. Greve, M. Kaufmann, P. Manolios, J S. Moore, S. Ray, J. L. Ruiz-Reina, R. Sumners, D. Vroon, and M. Wilding. Efficient Execution in an Automated Reasoning Environment. *Journal of Functional Programming*, To Appear.
[10] W. A. Hunt, Jr. The DE Language. In P. Manlolios, M. Kaufmann, and J S. Moore, editors, *Computer-Aided Reasoning: ACL2 Case Studies*, pages 119–131, Boston, MA, June 2000. Kluwer Academic Publishers.
[11] W. A. Hunt, Jr. and E. Reeber. Formalization of the DE2 Language. In W. Paul, editor, *Proceedings of the 13th Working Conference on Correct Hardware Design and Verification Methods (CHARME 2005)*, LNCS, Saarbrücken, Germany, 2005. Springer-Verlag.
[12] M. Kaufmann, P. Manolios, and J S. Moore. *Computer-Aided Reasoning: An Approach*. Kluwer Academic Publishers, Boston, MA, June 2000.
[13] M. Kaufmann and J S. Moore. Design Goals of ACL2. Technical Report 101, Computational Logic Incorporated (CLI), 1717 West Sixth Street, Suite 290, Austin, TX 78703, 1994.
[14] M. Kaufmann and J S. Moore. Structured Theory Development for a Mechanized Logic. *Journal of Automated Reasoning*, 26(2):161–203, 2001.
[15] M. Kaufmann and R. Sumners. Efficient Rewriting of Data Structures in ACL2. In D. Borrione, M. Kaufmann, and J S. Moore, editors, *Proceedings of 3rd International Workshop on the ACL2 Theorem Prover and Its Applications (ACL2 2002)*, pages 141–150, Grenoble, France, April 2002.
[16] H. Liu and J S. Moore. Executable JVM Model for Analytical Reasoning: A Study. In *ACM SIGPLAN 2003 Workshop on Interpreters, Virtual Machines, and Emulators*, San Diego, CA, June 2003.
[17] P. Manolios, K. Namjoshi, and R. Sumners. Linking Model-checking and Theorem-proving with Well-founded Bisimulations. In N. Halbwacha and D. Peled, editors, *Proceedings of the 11th International Conference on Computer-Aided Verification (CAV 1999)*, volume 1633 of *LNCS*, pages 369–379. Springer-Verlag, 1999.
[18] J S. Moore, T. Lynch, and M. Kaufmann. A Mechanically Checked Proof of the Kernel of the AMD5K86 Floating-point Division Algorithm. *IEEE Transactions on Computers*, 47(9):913–926, September 1998.
[19] S. Ray, J. Matthews, and M. Tuttle. Certifying Compositional Model Checking Algorithms in ACL2. In W. A. Hunt, Jr., M. Kaufmann, and J S. Moore, editors, *4th International Workshop on the ACL2 Theorem Prover and Its Applications (ACL2 2003)*, Boulder, CO, July 2003.
[20] D. Russinoff. A Mechanically Checked Proof of IEEE Compliance of a Register-Transfer-Level Specification of the AMD-K7 Floating-point Multiplication, Division, and Square Root Instructions. *LMS Journal of Computation and Mathematics*, 1:148–200, December 1998.
[21] D. M. Russinoff and A. FLatau. RTL Verification: A Floating Point Multiplier. In M. Kaufmann, P. Manolios, and J S. Moore, editors, *Computer-Aided Reasoning: ACL2 Case Studies*, pages 161–187. Kluer Academic Publishers, 2000.
[22] J. Sawada and W. A. Hunt, Jr. Trace Table Based Approach for Pipelined Microprocessor Verification. In O. Grumberg, editor, *Proceedings of the 9th International Conference on Computer-Aided Verification (CAV 1997)*, volume 1254 of *LNCS*, pages 364–375. Springer-Verlag, 1997.
[23] G. L Steele, Jr. *Common Lisp the Language*. Digital Press, 30 North Avenue, Burlington, MA 01803, 2nd edition, 1990.
[24] D. E. Thomas and P. R. Moorby. *The Verilog® Hardware Description Language*. Kluwer Academic Publishers, Boston, MA, 3rd edition, 1996.

Application of Lifting in Partial Design Analysis

Marc Herbstritt   Vanessa Struve   Bernd Becker

Albert-Ludwigs-University, Freiburg im Breisgau, Germany
{herbstri,struve,becker}@informatik.uni-freiburg.de

Abstract—In the past, we have investigated satisfiability-based combinational equivalence checking and bounded model checking of *partial circuit designs*, i.e., circuit designs where one or more components of the design are not implemented yet. Especially for satisfiability-based bounded model checking, typically a *counterexample* is generated in case that the specification property is violated. Minimizing the number of assigned variables in the satisfying assignment that corresponds to such a counterexample is the objective of *lifting*.

In this work we show that lifting is also feasible and profitable for counterexamples obtained via satisfiability-based bounded model checking of partial designs. We provide first experimental results on this issue that show its feasibility. Furthermore, we present a novel application scenario for lifting in the context of *automated blackbox synthesis*. This is a useful concept that can be applied during combinational equivalence checking of partial circuit designs, where realizability of the missing components was already proven, but the functionality of the missing components is still unknown.

As a summary, this work provides first experimental results as well as a novel concept regarding the application of lifting for the analysis of partial designs.

I. Introduction

A partial circuit design corresponds to an implementation of a circuit that is incomplete. The occurrence of such partial circuit designs is many-fold and a very intuitive case is, e.g., when several design teams are working together on a design that is still in an early phase and hence some components are not implemented yet. But also the abstraction of system modules can lead to partial circuit designs. The missing components are denoted as *blackboxes*.

In the past, the analysis of partial circuit designs has been investigated for several problem settings such as combinational equivalence checking (CEC) [1], [2], symbolic model checking (SMC) [3], and bounded model checking (BMC) [4], [5]. All these investigations had to provide an adequate modeling of the blackbox behavior, ranging from a conservative extension of propositional logic called 01X-logic to the application of universal quantifiers resulting, e.g., in QBF-formulas.

Meanwhile, simplification of counterexamples, especially in the context of satisfiability-based BMC, has become a vital task for strengthening the usability of formal verification tools. In [6], such a simplification technique - called *lifting* - was presented for counterexamples computed by BMC of *complete* circuit designs. The main purpose of lifting is to detect syntactically and semantically redundant variable assignments within the counterexample, so that is easier for a designer to interpret it. Put another way, a simplified counterexample focuses on the root sources of the error that is responsible for the specification property to fail.

In this work we (1) investigate how lifting can be applied to counterexamples that were computed by BMC of partial circuit designs, (2) present first experimental results showing that lifting is also profitable for such counterexamples, and (3) investigate a novel application of lifting that is rooted in the context of combinational equivalence checking of partial circuit designs, i.e., we show that lifting can be used for *automated completion* of partial circuit designs.

The paper is structured as follows. In the next section, we provide preliminaries. Then, in Section III we present our lifting strategies for counterexample in the context of partial circuit designs together with an experimental evaluation of these strategies. In Section IV, the concept for using lifting for automated blackbox synthesis is described. Finally, we conclude the paper in Section V.

II. Preliminaries

We give some preliminaries for the SAT-based analysis of partial circuit designs that relies on 01X-logic as well as lifting for the simplification of counterexamples..

A. 01X-Logic

For handling the unknown behavior of the blackboxes, a third logical value X can be used, meaning that a signal is either 0 or 1 but unknown. Such 01X-logic is well established in ATPG, see [7] for details. Relying on the fact that 01X-logic conservatively extends propositional logic, e.g., $\text{AND}(1,X) = X$, we make use of a two-valued encoding as suggested in [8]. There, the three logical values $0, 1$, and X are encoded as

$$0_{01X} := (1,0)$$
$$1_{01X} := (0,1)$$
$$X_{01X} := (0,0).$$

The basic boolean operators AND, OR, and NOT can be adapted as follows ((a_0, a_1) and (b_0, b_1) are tuples for the 01X-variables):

$$\text{AND}_{01X}((a_0,a_1),(b_0,b_1)) := (a_0 + b_0, a_1 \cdot b_1)$$
$$\text{OR}_{01X}((a_0,a_1),(b_0,b_1)) := (a_0 \cdot b_0, a_0 + b_0)$$
$$\text{NOT}_{01X}((a_0,a_1)) := (a_1, a_0).$$

This encoding technique allows to transform problems derived from partial circuit designs where the blackbox behavior

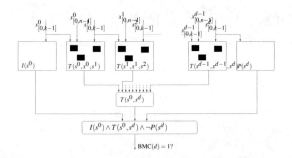

Fig. 1: BMC of blackbox designs.

Algorithm 1 Brute-Force-Lifting for Complete Designs

1: **procedure** BRUTEFORCELIFTING($\hat{F}, o, A$)
2:    $F'' \leftarrow$ substitute o with $\bar{o}$ in $\hat{F}$
3:    **for each**(literal $l \in A$)
4:      $F' \leftarrow F'' \wedge \bigwedge(A \setminus l)$
5:      **if** (SATSOLVE(F') $\neq$ SATISFIABLE) **then**
6:        $A \leftarrow A \setminus l$
7:      **end if**
8:    **end for**
9:    **return** A
10: **end procedure**

is modeled using 01X-logic, as sketched in Fig. 1 for BMC, into a pure propositional SAT problem that can be solved using off-the-shelf SAT-solvers, e.g., a CNF SAT-solver [9], a sequential SAT-solver [10], or a structural SAT-solver [11].

In our work here, we make use of our own implementation of a structural SAT-solver in the style of [11] for which we made extensions towards BMC of blackbox designs [12], [4].

B. Lifting

Lifting is a technique for removing assignments from a SAT-solution such that the initial objective is still assured. In [6], among others, a brute-force approach is presented that is used for lifting counterexamples found by SAT-based BMC. Put another way, the BMC-formula

$$BMC(k) = I(s_0) \cdot \bigwedge_{i=0}^{(k-1)} T(s_i, x_i, s_{(i+1)}) \cdot \overline{P(s_k)} \quad (1)$$

is transformed into a CNF and a SAT-solver is applied to find a satisfying assignment that corresponds to a counterexample (CE) that falsifies property P. Such a counterexample is a path $(s_0, s_1, \ldots, s_k)$ of length k that starts in an initial state s_0 and leads to a state s_k where property P is violated. The basic lifting algorithm as suggested in [6] is given in Algorithm 1. The main idea is that there is an objective o, a set A of candidate variables for lifting, and a modified formula $\hat{F}$ with the following intention. The objective o corresponds to property P in our case, A contains the values of the satisfying assignment for *interesting* variables, e.g., the primary inputs x_i and the state variables s_i but not for the Tseitin-variables that were used for CNF-encoding, and $\hat{F} := \bigwedge_{i=0}^{(k-1)} T(s_i, x_i, s_{(i+1)}) \cdot \overline{P(s_k)}$ to enable lifting of initial state literals.

The algorithm computes correct results, i.e., the reduced assignment $(A \setminus l)$ is still a counterexample. This is because when formula F' in the above algorithm is unsatisfiable, then it is not possible to reach a state s_k where property P holds (this is stated by replacing o by $\bar{o}$ in F''), hence property P is still violated – for all values of the removed literal l.

Another interesting point is that the result of the lifting procedure depends on the variable order. For a simple example, assume the following formula

$$(a \vee b) \wedge (a \vee c \vee d) \wedge (\bar{a} \vee b \vee c) \quad (2)$$

for which $\{a, b, c, d\}$ is a satisfying assignment. Now, if the variable order for lifting is $a < b < c < d$, then $\{b, c\}$ is the minimal assignment. But when the order is $b < a < c < d$, then $\{a, c, d\}$ is the result of lifting.

III. LIFTING IN THE CONTEXT OF BOUNDED MODEL CHECKING

For checking how lifting can be applied in the context of BMC of partial circuit designs, we assume in a first step that the blackboxes are modeled using 01X-logic and that we apply the two-valued encoding as presented in Sect. II.

A. Lifting Strategies

We propose three lifting strategies for removing redundant variable assignments from the counterexample that take into account the two-valued encoding used for handling 01X-logic. As such, the strategies not only remove redundant variable assignments syntactically but also semantically.

   a) Pair Lifting (PL): Since a 01X-variable a is encoded by a tuple (a_0, a_1), *pair lifting* tries to lift both variables a_0 and a_1 in parallel. Only if both variables can be lifted, then $a = (a_0, a_1)$ is removed from the counterexample. If a cannot be lifted, it remains as fully specified 01X-value in the counterexample.

   b) Pair-Split Lifting (PSL): In contrast to pair lifting, *pair-split lifting* allows also to only lift a_0 or a_1 from the tuple (a_0, a_1). Such a bisection of a 01X-value leads to the information that, e.g., a 01X-variable a can be 0 or X, but definitely not 1. This can ease the interpretation of a counterexample.

   c) Single-Encoding-Variable Lifting (SEVL): For partial satisfying assignments, it may also be the case that for a tuple (a_0, a_1) only one of the two variables has got a value assigned. By intuition, such a variable may then be a good candidate for lifting, since it would result in lifting the whole 01X-variable $a = (a_0, a_1)$. Such a *single-encoding-variable lifting* may be used as preprocessing.

B. Preliminary Experiments

We have implemented the above described lifting strategies into our bounded model checking framework as described in [12], [4], [5]. As benchmark example, we computed a counterexample for an erroneous design of a VLIW ALU,

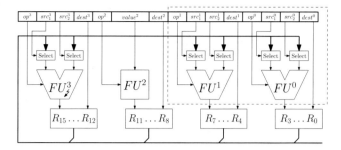

Fig. 2: VLIW ALU with an error wrt. the XOR-implementation in functional unit FU_3.

width	# var.	# ass.	PL	PSL	SEVL
2	756	686	616	650	6
4	1124	1054	984	1018	6
16	3332	3262	3190	3226	6
24	4804	4734	4662	4698	6
32	6276	6206	6134	6170	6
40	7748	7678	7606	7742	6
48	9220	9150	9078	9114	6

TABLE I: Results for lifting strategies.

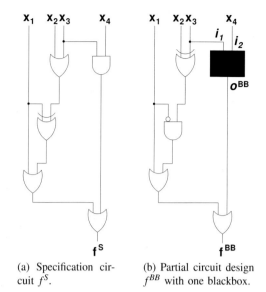

(a) Specification circuit f^S.

(b) Partial circuit design f^{BB} with one blackbox.

Fig. 3: Specification circuit for f^S and partial circuit design f^{BB}.

as depicted in Fig.2. The VLIW ALU consists of 4 functional units whereby the fourth unit has an error due to an incorrect implementation of the XOR function (the OR function is computed instead!). A more detailed description of the VLIW ALU can be found in [13], [14]. The VLIW ALU is configurable in its word width, thus enabling to scale the complexity for the underlying decision procedure. To get blackbox designs, we removed two functional units for which we were aware that they are unnecessary for the computation of the counterexample, and replaced them by blackboxes.

Table I gives results for the VLIW ALU for different word width. Column 2 gives the number of variables and column 3 the number of assigned variables of the partial satisfying assignment generated by the structural SAT solver. Columns 4, 5, and 6 contain the number of liftable variables according to the *pair lifting* (PL), *pair-split lifting* (PSL) and *single-encoding-variable lifting* (SEVL) strategy, resp.

As can been, PL and PSL are able to remove a large number of assigned variables, especially for *pair lifting* it becomes clear that complete 01X-values are liftable from the counterexample. When one favors a minimal 01X-counterexample, *pair-split lifting* has to be preferred since it removes again a considerable amount of variables compared to *pair lifting*. Unfortunately, the *single-encoding-variable lifting* strategy does not really pay off. As a summary, both *pair lifting* and *pair-split lifting* are suitable methods to simplify a counterexample that stems from bounded model checking of partial circuit designs.

IV. LIFTING IN THE CONTEXT OF AUTOMATED BLACKBOX SYNTHESIS

In this section we are investigating how a blackbox can be synthesized in case that a partial circuit design has been proven to be realizable, i.e., the missing components can be completed such that it becomes equal to the specification. This application is located in the context of combinational equivalence checking of partial circuit designs.

A. An Example

Assume a specification circuit f^S and a partial circuit design f^{BB}, as depicted in Figure 3a and 3b. The corresponding truth tables are given in Figure IV-A whereby 01X-logic is used for modeling the blackbox behavior of f^{BB}.

The truth tables, using 01X-logic for modeling the blackbox behavior, are as follows:

x_1	x_2	x_3	x_4	f^S	f^{BB} (01X)
0	0	0	0	0	X
0	0	0	1	0	X
0	0	1	0	1	1
0	0	1	1	1	1
0	1	0	0	1	1
0	1	0	1	1	1
0	1	1	0	1	X
0	1	1	1	1	X
1	0	0	0	1	1
1	0	0	1	1	1
1	0	1	0	1	1
1	0	1	1	1	1
1	1	0	0	1	1
1	1	0	1	1	1
1	1	1	0	1	1
1	1	1	1	1	1

$$\forall i_1 \, \forall i_2 \, \exists o^{BB} : cond'$$
$$= \quad \forall i_1 \, \forall i_2 \, \exists o^{BB} : \quad \forall x_1, \ldots, x_4 : \quad \overline{(H(x_1, x_2, x_3, x_4, i_1, i_2) + cond(x_1, x_2, x_3, x_4, o^{BB}))}$$
$$= \quad \forall i_1 \, \forall i_2 \, \exists o^{BB} : \quad \forall x_1, \ldots, x_4 : \quad \overline{((i_1 \equiv h^1(x_1, x_2, x_3, x_4)) \cdot (h^2(x_1, x_2, x_3, x_4)) + cond(x_1, x_2, x_3, x_4, o^{BB}))}$$
$$= \quad \forall i_1 \, \forall i_2 \, \exists o^{BB} : \quad \forall x_1, \ldots, x_4 : \quad \overline{((i_1 \equiv x_3) \cdot (i_2 \equiv x_4) + (f^S(x_1, x_2, x_3, x_4) \equiv f^{BB}(x_1, x_2, x_3, x_4, o^{BB})))}$$
$$= \quad \forall i_1 \, \forall i_2 \, \exists o^{BB} : \quad \forall x_1, \ldots, x_4 : \quad \Big(((i_1 \equiv x_3) \cdot (i_2 \equiv x_4)) \to (f^S(x_1, x_2, x_3, x_4) \equiv f^{BB}(x_1, x_2, x_3, x_4, o^{BB})) \Big)$$

Fig. 4: QBF formula for the *input-exact-check*.

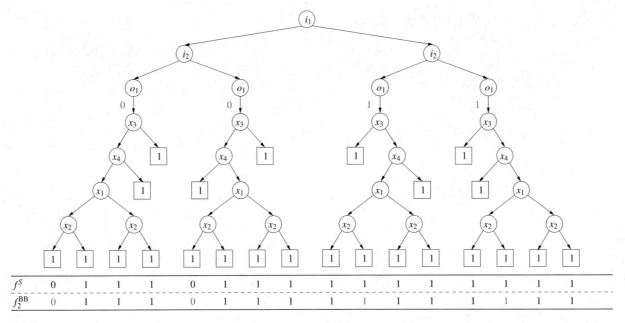

Fig. 5: AND/OR-tree for QBF formula of Fig. 4.

According to the work of Scholl and Becker in [1], analyzing equivalence between f^S and f^{BB} can be done in an exact manner by using the *input-exact-check* of [1]. Using the terminology of [1], we have to check whether the QBF formula of Fig. 4 is valid. The validity check can be performed by a QBF-solver, e.g., [15], [16], [17].

The formula of Fig. 4 is indeed valid which can be understood by looking at the AND/OR-tree of Figure 5. There, a left (right) edge corresponds to assigning value 0 (1) to the variable attached to the vertex from which the edge emanates. For the existentially quantified variable o^{BB}, there is only one outgoing edge, indicating that some value has to be found such that all leaves below the sub-tree of o^{BB} result in value 1. This has to be done especially for the four cases where a value X is computed by f^{BB} (see the table in Figure IV-A): $(0,0,0,0)$, $(0,0,0,1)$, $(0,1,1,0)$, and $(0,1,1,1)$. For the input assignments $(0,0,0,0)$ and $(0,0,0,1)$ this is achieved by setting o^{BB} to 0, and for the input assignments $(0,1,1,0)$ and $(0,1,1,1)$ by setting o^{BB} to 1.

When a QBF-solver proves that the formula of Fig. 4 is valid, then there exists a completion f^{BB}_{complete} of f^{BB} that is equal to f^S, i.e., $\forall x_1, x_2, x_3, x_4 : f^S(x_1, x_2, x_3, x_4) = f^{BB}_{\text{complete}}(x_1, x_2, x_3, x_4)$.

The problem that remains is that the blackbox is typically embedded in some existing partial circuit design and hence the task is to find a suitable implementation of the blackbox component, i.e., at first we have to find the function $f^{BB}_{\text{sub}}(i_1, i_2)$ such that $f^{BB}_{\text{complete}} := f^{BB}(x_1, x_2, x_3, x_4, o^{BB})|_{o^{BB} \leftarrow f^{BB}_{\text{sub}}(i_1, i_2)}$. For our example, the function $f^{BB}_{\text{sub}}(i_1, i_2)$ is

i_1	i_2	$f^{BB}_{\text{sub}}(i_1, i_2)$
0	0	0
0	1	0
1	0	1
1	1	1

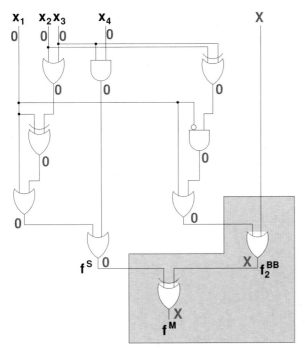

Fig. 6: Miter for f^S and f^{BB} using the X-value for modeling the blackbox output.

and can be implemented by using $f_{\text{sub}}^{BB}(i_1, i_2) := x_3$.

This function is indeed computed *implicitly* by the QBF-solver, however, it cannot be extracted in a straightforward manner. We now suggest a different approach that is based on 01X-logic and lifting of satisfying assignments.

B. Automated Blackbox Synthesis

Assume that we know that f^{BB} is realizable. This means that the output of the miter depicted in Figure 6, that relies on 01X-logic for modeling the blackbox output, is *never* equal to 1. In terms of 01X-logic, it means that the output is never 1_{01X} (i.e., $\neq 1_{01X}$) which in turn means that it is always either 0_{01X} or X_{01X}. The former case corresponds to input assignments $x^\equiv$ where $f^S(x^\equiv) = f^{BB}(x^\equiv)$, i.e., function f^{BB} computes a concrete boolean value 0 or 1 that is equal to the value computed by f^S. The latter case corresponds to input assignment $x^?$ for which function $f^{BB}(x^?) = X$ and thus the equality with respect to $f^S(x^?)$ is unclear. But keep in mind that we already know that f^{BB} is realizable, and hence we know that there exists an assignment to o^{BB} such that $f^{BB}(x^?)$ is equal to the value computed by f^S. Regarding our task of blackbox synthesis, i.e., to identify the function $f_{\text{sub}}^{BB}(i_1, i_2)$ in our example above, we are interested only in the input assignments $x^?$. For the other input assignments $x^\equiv$ the output of f_{sub}^{BB} has no impact and thus can be considered as don't care assignments that may be useful for further optimization of the blackbox implementation. Our focus is on identifying the mandatory values of f_{sub}^{BB} for the input assignments $x^?$. For the desired identification, we suggest the following approach:

1) We build a miter of f^S and f^{BB} by using 01X-logic, i.e., the two-valued encoding scheme of Jain et al. is used. This results in tuples for (f_0^S, f_1^S) for the specification f^S, (f_0^{BB}, f_1^{BB}) for the blackbox design f^{BB}, and (f_0^M, f_1^M) for the miter output. Blackbox outputs are managed by using tuples (o_0^{BB}, o_1^{BB}) for each blackbox output. This encoded circuit can be transformed into a CNF φ.

2) We make use of an assumption list α given to the SAT-solver telling him to assign some variables to some specific value already before the search. This is especially useful for stating that the blackbox outputs are assigned to X, i.e., $\alpha = (o_0^{BB} = 0, o_1^{BB} = 0)$. Many SAT-solvers, e.g., MiniSat [9], support such an assumption list as additional input.

3) By adding the constraint $(\overline{f_0^M} \cdot \overline{f_1^M})$ to CNF φ, a SAT-solver finds solutions that correspond to input assignments $x^?$, i.e., the value X_{01X} is produced at the miter output. Put another way, we call a SAT-solver $\mathcal{S}$ with $\mathcal{S}(\alpha, \varphi \cup \{\overline{f_0^M}\} \cup \{\overline{f_1^M}\})$.

4) Let $\gamma_X^?$ be a satisfying assignment found by $\mathcal{S}$. Such a scenario is depicted for the input assignment $(0,0,0,0)$ in Figure 6 (for the non-encoded miter circuit). Figure 7 shows the excerpt for this scenario of the grey-shaded area in Figure 6, but now for the two-valued encoding of the miter.

5) It holds that $\gamma_X^?$ is also a satisfying solution for $\varphi \cup \{\overline{f_1^M}\}$ under the assumption of assignments α, since it only weakens the previous constraint added to φ.

6) Considering only $\overline{f_1^M}$ as additional constraint means that we now allow also the value 0_{01X} for the miter output. This corresponds to the case where f^S and f^{BB} compute the same *boolean* value 0 or 1.

7) We are now able to lift either o_0^{BB} or o_1^{BB}, but not both, from the satisfying assignment $\gamma_X^?$: $\gamma_0^? := \gamma_X^? \setminus \overline{o_0^{BB}}$ and $\gamma_1^? := \gamma_X^? \setminus \overline{o_1^{BB}}$. It holds that exactly one of both assignments still satisfies $\varphi \cup \{\overline{f_1^M}\}$. This is because when one assumes that both o_0^{BB} and o_1^{BB} are liftable, then no X-value would be present in the encoded circuit and hence no assignments $x^?$ would be found. When $\overline{o_0^{BB}}$ is liftable, then the output value of the blackbox function has to be 0, and when $\overline{o_1^{BB}}$ is liftable, then it has to be 1. This way we are able to identify values for the function, e.g., for our example above it must hold $f_{\text{sub}}^{BB}(0,0) = 0$, since $\overline{o_0^{BB}}$ is liftable. At this point, our lifting strategy *pair-split lifting* from the previous section is profitable.

8) W.l.o.g., let $\gamma_0^?$ be the assignment that satisfies $\varphi \cup \{\overline{f_1^M}\}$. $\gamma_0^?$ is also related to a fully specified assignment to the inputs $x_1, x_2, \ldots, x_n$ of the circuit. The corresponding values of $x_1, x_2, \ldots, x_n$ are either 0 or 1 but not X. Now we can again apply lifting to potentially lift an assignment to variable x_i to the value X. This means that both $(x_1, \ldots, x_{i-1}, 0, x_{i+1}, \ldots, x_n)$ and $(x_1, \ldots, x_{i-1}, 1, x_{i+1}, \ldots, x_n)$ are input assignments for which the *same* value of (o_0^{BB}, o_1^{BB}) can be used. In our example this is $(o_0^{BB}, o_1^{BB}) = (-, 0)$, since o_0^{BB} was already lifted. Thus, for our example, we are able to

identify that for the input assignments $(0,0,0,0)$ and $(0,0,0,1)$ the blackbox function f^{BB}_{sub} has to output the value 0.

9) It may be possible to lift more than one assignment regarding the assignment of the primary inputs. Hence, we may apply our lifting procedure iteratively by assuring that our objective $\overline{f^M_1}$ still holds.

As a summary, the above described scheme does the following:

1) Find an assignment $x^? = (x_1, x_2, x_3, x_4)$ for which $f^{BB}(x^?) = X_{01X}$.
2) Weaken the satisfiability constraint to allow $f^M(x^?) \in \{0_{01X}, X_{01X}\}$.
3) Find a concrete boolean value c for o^{BB} such that $f^M|_{o^{BB} \leftarrow c}(x^?) = 0$, i.e., the substitution of c for o^{BB} makes f^{BB} equal to f^S for $x^?$.
4) Try to lift variables from $x^?$ such that some x_i can have value X_{01X}. Let's denote by $x^{?,ext}$ such a lifted assignment. For this assignment it also holds that $f^M|_{o^{BB} \leftarrow c}(x^{?,ext}) = 0$, meaning that f^{BB} is equal to f^S when using value c for o^{BB}.
Iterate this step.

V. CONCLUSIONS

We have investigated the application of lifting for problems derived from blackbox designs. We proposed dedicated lifting strategies that experimentally proved to be beneficial for counterexamples that were computed by BMC of partial circuit designs. Furthermore, we proposed the usage of lifting to solve the problem of automated blackbox synthesis for partial circuit designs that were proved to be realizable.

ACKNOWLEDGMENTS

We'd like to thank Tobias Nopper for providing the VLIW ALU example and Christoph Scholl for fruitful discussion, especially regarding the input-exact-check of [1].

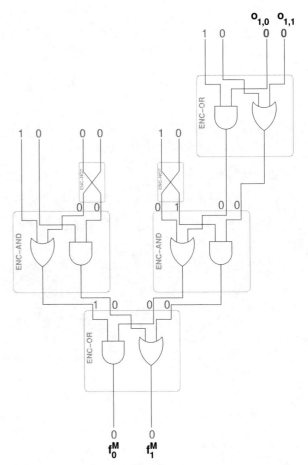

Fig. 7: Excerpt for the two-valued encoded miter for f^S and f^{BB} regarding the grey-shaded area of Figure 6.

REFERENCES

[1] C. Scholl and B. Becker, "Checking equivalence for partial implementations," in *Proc. of Design Automation Conference (DAC)*, 2001, pp. 238–243.
[2] ——, "Checking equivalence for circuits containing incompletely specified boxes," in *Proc. of Int'l Conf. on Comp. Design (ICCD)*, 2002.
[3] T. Nopper and C. Scholl, "Approximate symbolic model checking for incomplete designs," in *Proc. of Int'l Conf. on Formal Methods in Computer-Aided Design (FMCAD)*, 2004, pp. 290–305.
[4] M. Herbstritt, B. Becker, and C. Scholl, "Advanced SAT-techniques for bounded model checking of blackbox designs," in *Proc. of Int'l Workshop on Microprocessor Test and Verification (MTV)*, Austin (TX), USA, 2006, pp. 37–44.
[5] M. Herbstritt and B. Becker, "On Combining 01X-Logic and QBF," in *Proc. of Int'l Conf. on Computer-Aided Systems Theory (EuroCAST)*, ser. LNCS, 2007, in press.
[6] K. Ravi and F. Somenzi, "Minimal assignments for bounded model checking." in *Prof. of Int'l Conf. on Tools and Algorithms for the Construction and Analysis of Systems (TACAS)*, ser. LNCS, 2004, pp. 31–45.
[7] M. Abramovici, M. Breuer, and A. Friedman, *Digital Systems Testing and Testable Design*. Computer Science Press, 1990.
[8] A. Jain et al., "Testing, Verification, and Diagnosis in the Presence of Unknowns," in *Proc. of VTS'00*, pp. 263–269.
[9] N. Eén and N. Sörensson, "An extensible sat-solver." in *Prof. of Int'l Conf. on Theory and Applications of Satisfiability Testing (SAT)*, ser. LNCS, 2003, pp. 502–518.
[10] G. Parthasarathy et al., "Safety Property Verification Using Sequential SAT and Bounded Model Checking." *IEEE Design & Test of Computers*, vol. 21, no. 2, pp. 132–143, 2004.
[11] A. Kuehlmann et al., "Robust Boolean Reasoning for Equivalence Checking and Functional Property Verification," *IEEE Trans. on CAD*, vol. 21, no. 12, pp. 1377–1394, 2002.
[12] M. Herbstritt and B. Becker, "On SAT-based Bounded Invariant Checking of Blackbox Designs," in *Proc. of Int'l Workshop on Microprocessor Test and Verification (MTV)*, Austin (TX), USA, pp. 23–28.
[13] T. Nopper, C. Scholl, and B. Becker, "Computation of minimal counterexamples by using black box techniques and symbolic methods," in *Proc. of Int'l Conf. on Computer-Aided Design (ICCAD)*, 2007, in press.
[14] SFB/TR 14 AVACS - Subproject S1, "The VLIW ALU Benchmark," http://www.avacs.org, 2007.
[15] E. Giunchiglia, M. Narizzano, and A. Tacchella, "Quantifier structure in search based procedures for QBFs." in *Proc. of Conference on Design, Automation and Test in Europe (DATE)*, 2006, pp. 812–817.
[16] M. Benedetti, "Evaluating QBFs via Symbolic Skolemization." in *Proc. of Int'l Conf. on Logic for Programming, Artificial Intelligence, and Reasoning (LPAR)*, Montevideo, Uruguay, 2004, pp. 285–300.
[17] A. Biere, "Resolve and expand." in *Prof. of Int'l Conf. on Theory and Applications of Satisfiability Testing (SAT)*, Vancouver, BC, Canada, 2004.

Model Checking Bluespec Specified Hardware Designs

Gaurav Singh and Sandeep K. Shukla,
FERMAT Lab,
Virginia Polytechnic and State University, Blacksburg, VA.
{gasingh,shukla}@vt.edu

Abstract

Using RTL (Register Transfer Level) models for the verification of complex hardware designs involves reducing the state space of designs using various abstraction techniques. In this paper, we propose faster and earlier verification of hardware designs at a level of abstraction above RTL. We consider a high-level (above RTL) hardware model that uses atomic rules to describe the behavior of a design, which can then be synthesized to RTL code. Bluespec System Verilog (BSV) is an example of such a high-level specification language. We propose a methodology for verification of BSV models using Spin, which is a Model Checking tool. Verification of high-level BSV models may avoid the need for using abstraction techniques since such models already ignore various low-level details that are irrelevant for verifying a design's behavioral properties. Moreover, using our proposed methodology different behaviors of BSV models can be efficiently verified at high-level aiding in faster verification.

1 Introduction

Traditionally, hardware designs are described at RTL. Formal verification of complex hardware designs at RTL is known to be associated with the state space explosion problem. To resolve this, various abstraction techniques are employed in order to generate abstract models of designs having reduced state space. Such abstract models are then used to verify the desired correctness properties of hardware designs. Recently, hardware synthesis has been proposed from high-level (above RTL) models of hardware designs. Such models are used to describe the behavior of the designs at a high-level, and can be automatically converted into RTL descriptions using various high-level synthesis tools.

High-level models of a hardware design can be used for faster verification of its desired behavioral properties early in the design cycle. This is because high-level description of a hardware design is usually more concise than its corresponding RTL code since high-level models ignore various details irrelevant to the behavior of a design. In most cases, this translates to reduced state space of the design which helps in faster verification at high-level.

In this paper, we propose the formal verification of high-level *Bluespec System Verilog (BSV)* [1, 2] description of a hardware design using *Spin*, which is a Model Checking tool used for the verification of concurrent systems [3, 4]. *Spin* supports verification of correctness properties expressed as assertions, Linear Temporal Logic (LTL), etc. For verification using *Spin*, a system should be described in terms of interacting processes using *PROMELA*, an input specification language for *Spin*. We present a methodology which involves the conversion of rule-based *BSV* description of a hardware design into a corresponding process-based model in *PROMELA*, such that the behavior of the *PROMELA* model is same as that of the hardware generated using *BSV*. Such a model can then be automatically verified using *Spin*.

The paper is organized as follows. In Section 2, we explain how hardware designs can be synthesized from *Bluespec System Verilog (BSV)*. Section 3 provides a small introduction of Model Checking tool *Spin* and its input specification language *PROMELA*. Section 4 shows how a rule-based *BSV* description of a hardware design can be translated into process-based *PROMELA* description for its verification. Section 5 presents the main steps of our proposed methodology for the verification of high-level *BSV* models. Finally, Section 6 concludes the paper with a brief discussion.

2 Synthesis from BSV

Most existing high-level synthesis techniques convert the high-level (above RTL) input specifications of hardware designs into intermediate representations in the form of CDFGs (Control Data Flow Graphs) or other variants of CDFGs such as MTGs (Multi Threaded Graphs) [5]. These intermediate representations are then converted into corresponding RTL code. Most high-level synthesis tools are based on generating RTL description of a hardware design from its high-level C-based behavioral specification.

In this work, we consider a different model where the behavior of the target hardware is described at high-level in terms of atomic rules. Such a model can be expressed using *Bluespec System Verilog (BSV)*, a high-level hardware specification language. In *BSV*, each rule consists of a set of atomic operations (called its *body*) which can be executed under a given condition called the *guard* of the rule. All operations in the body of a rule are executed atomically when its *guard* evaluates to *True* [1, 2].

Figure 1 shows the example of an abstract rule-based description of a simple *Packet Processor* written in *BSV*. rule A corresponds to the operations that the processor executes when a packet arrives. Then, operations in *rule B* process that packet. And finally, *rule C* corresponds to the departure of the packet.

```
// Receive packet.
rule A (True) ;
      x := x + 1 ;
endrule


// Process packet.
rule B (x > 0) ;
      x := x - 1 ;
      y := y + 1 ;
endrule


// Send packet.
rule C (y > 0) ;
      y := y - 1 ;
endrule
```

Figure 1. BSV Description of Packet Processor.

Such a rule-based *BSV* model then undergoes synthesis to generate RTL code [1, 2, 6]. Hardware synthesis from rule-based models can be achieved by implementing each *guard* and *body* as a combinational logic and synthesizing a control circuitry for appropriate scheduling and data-selection. Figure 2 shows hardware synthesized from a rule-based *BSV* description of a design. During the synthesis process, various concurrency and synchronization issues are automatically handled by the high-level synthesis tool. *Bluespec Compiler* is an example of such a synthesis tool which converts a *BSV* specification of a hardware design into RTL description.

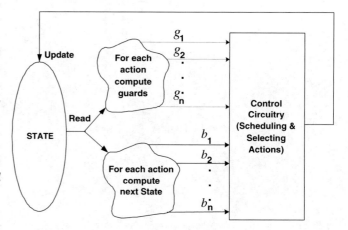

Figure 2. Synthesis from Concurrent Rules.

In the generated hardware, maximal set of non-conflicting rules are executed concurrently in a clock cycle as long as their concurrent execution corresponds (matches in output) to a sequential execution where only one rule is executed per clock cycle. Thus, for the *Packet Processor* design shown in Figure 1, *rule A* and *rule C* can execute concurrently in the same clock cycle when their guards evaluate to *True*. This is because concurrent execution of *rule A* and *rule C* corresponds to sequentially executing *rule C* in first clock cycle followed by *rule A* in the next clock cycle.

An example of conflict is two rules updating the same register. In Figure 1, *rule A* and *rule B* conflict with each other because they both update the same state element (register) x. In case of a conflict, higher priority rule is chosen for execution. For example, in the *Packet Processor* design, *rule B* is always chosen over *rule A* whenever both are enabled (guards evaluated to *True*) in the same clock cycle.

3 Spin Model Checker

Spin is a Model Checking tool used to verify the correctness requirements of software systems [3, 4]. The system to be verified is described in terms of concurrent processes using *PROMELA*, which is a input specification language of *Spin*.

Each process consists of data declarations and executable statements. An example of a process called

Proc modeled described in *PROMELA* is shown in Figure 3, where *state* and *tmp* are the global and local state variables respectively. When $(state == 1)$ holds *True*, variable *tmp* is assigned the value $state+1$. Multiple such processes can be described to model the functionality of a system in *Spin*.

```
byte state = 1;
proctype Proc()
{
    byte tmp;
    (state == 1) → tmp = state + 1;
}
```

Figure 3. Example Process modeled in PROMELA

Spin Model Checker also allows the specification of the desired properties of a system in terms of assertions as well as Linear Temporal Logic (LTL), which can then be used by *Spin* to verify the correctness of the system. For verification, all the possible combinations of execution of different processes are checked. Various statements of different processes of a system are interleaved among each other for exploring all the possible behaviors.

4 BSV to PROMELA

As mentioned earlier, the rule-based *BSV* model of a hardware design can be exploited for faster and early verification of the correctness properties of the design. The atomicity of the rules enforce the requirement that all the operations of a particular rule execute together without being interleaved or interrupted by operations of any other rule. This reduces the number of different behaviors of the design, thus expediting the verification process. We propose a methodology which uses the high-level description of a hardware design expressed in terms of atomic rules for faster verification.

As part of our proposed approach, we first convert the concurrent rule-based description of a design to a concurrent process-based *PROMELA* description. Figure 4 shows the converted *PROMELA* description of the *Packet Processor* design. The following steps are involved in the translation of a *BSV* description into *PROMELA* code -

1. A rule in *BSV* consists of a set of operations of a design which are executed atomically when the guard of the rule evaluates to *True* in a clock cycle. As shown in Figure 4, we model each rule of the *Packet Processor* design (shown in Figure 1) as

```
proctype processC()
{
    do
    :: atomic { { p == 1 → p = 2 ; }
                unless
                { ( p == 1 && y > 0 ) → { y = y - 1 ;
                                          p = 2 ; }
                }
              }
    od
}
proctype processB()
{
    do
    :: atomic { { p == 2 → p = 3 ; }
                unless
                { ( p == 2 && y ≤ 0 && x > 0 ) →
                                        { x = x - 1 ;
                                          y = y + 1 ;
                                          p = 3 ; }
                }
              }
    od
}
proctype processA()
{
    do
    :: atomic { { p == 3 → p = 1 ; }
                unless
                { ( p == 3 && ( y > 0 || x ≤ 0 ) ) →
                                        { x = x + 1 ;
                                          p = 1 ; }
                }
              }
    od
}
```

Figure 4. Process-based PROMELA Description of Packet Processor.

a process in *PROMELA*. *processA*, *processB* and *processC* correspond to *rule A*, *rule B* and *rule C* respectively. Each process consists of the operations of the corresponding rule and its guard.

2. Atomicity of rules is an important concept in *BSV*. We use *PROMELA* construct "atomic" to model such atomicity of operations [3]. This avoids the interleaving of operations of various processes, which is consistent with *BSV* semantics and aids in faster verification.

3. While performing the verification of a system, *Spin* considers all the possible behaviors of the concurrently executing processes. However, in rule-based synthesis, the scheduling of various rules (which is

done automatically by the synthesis tool) restricts the set of possible behaviors of the design. We exploit this observation to perform a faster verification of hardware designs using *Spin*. This is done by adding the scheduling and conflict information to the process-based description of the design.

The concurrent behavior of rules corresponds to a sequential ordering of their execution. During the synthesis process, such a sequential ordering is automatically generated by the rule-based synthesis tool. For the *Packet Processor*, the concurrent execution of the rules corresponds to the following sequential ordering - *rule C*, *rule B*, *rule A*. In Figure 4, variable p is used to execute the processes according to this sequential ordering in order to model the real hardware behavior. In this way, irrelevant behaviors of the design are ignored while analyzing it for verification purposes. For *PROMELA* model of Figure 4, variable p is initialized to 1. Thus, as per the sequential ordering, various processes will be executed in order. Starting with 1, value of p is changed to 2 and then to 3. This corresponds to one clock cycle of hardware execution.

Also note that operations of each rule are executed only when the guard of the corresponding rule evaluates to *True*. Otherwise, value of p is updated and next process is checked for execution. As shown in Figure 4, *PROMELA* construct "unless" is used to model such execution [3]. Additionally, conflicts between various rules also need to be modeled in the *PROMELA* description. This is required so that in one cycle processes corresponding to non-conflicting rules are only executed. For the *Packet Processor* design, *rule B* and *rule C* conflict with each other with *rule C* having priority. To model this, negation of guard of *processC* ($y \leq 0$) is added to the guard of *processB*. Conflict between *rule A* and *rule B* is modeled similarly.

4. Finally, "do", which is a repetition construct in *PROMELA* [3], is used for forwarding the execution of processes as in the hardware (cycle by cycle).

5 Verification Methodology

The main steps of our proposed methodology for the verification of *BSV* designs are -

1. Convert the rule-based *BSV* description of hardware design into process-based *PROMELA* description as discussed in Section 4. This includes the insertion of appropriate scheduling and conflict logic into *PROMELA* description to model the desired hardware behavior.

2. Express the desired properties of the design as Assertions, Linear Temporal Logic (LTL), Never claims, etc. and use the *Spin* Model Checker for verifying the properties of the converted *PROMELA* description [3].

Note that during synthesis of hardware designs, various scheduling and conflict information is generated by the rule-based synthesis tool based on the compile-time analysis of a *BSV* description. Thus, the process-based *PROMELA* model (similar to that shown in Figure 4) can also be automatically generated by a rule-based synthesis tool like *Bluespec Compiler*. The process-based model can then be given as an input to the *Spin* Model Checking tool for verification of the desired properties.

Example Properties

We used the proposed approach to verify the correctness of various realistic hardware designs. Below, we provide examples of two design properties we verified using our approach.

1. For the *Packet Processor* shown in Figure 1, a desirable property is to make sure that no new packets are received by the *Packet Processor* unless earlier packets have been processed. In terms of the state elements of the design, this property is expressed as: $assert\ (x < 2)$ [3]. *Spin* is then be used to verify the correctness of this property. Other desired properties of the design are verified similarly.

2. We also verified the behavior of a *Vending Machine* design. For this, *BSV* model of the *Vending Machine* design is converted into a *Spin* model. A example desired property of the *Vending Machine* is that a selected item is dispensed by the machine when sufficient amount of money has been inserted and no request for return of money is entered. All such properties of the design are modeled as LTL formulae and successfully verified using *Spin*.

6 Conclusion

In this paper, we proposed a methodology from verifying high-level *BSV* descriptions of hardware designs using *Spin* Model Checker. We claim that *Spin* is suitable for verification of *BSV* designs because of the following reasons -

1. *PROMELA* supports the description of concurrent systems at the desired level of abstraction. As shown in this paper, high-level *BSV* designs can be easily modeled using various *PROMELA* constructs.

2. Atomicity of *BSV* rules can be easily modeled using the "atomic' keyword of *PROMELA* [3].

3. Priority of operations and the execution of hardware designs can be efficiently modeled using *PROMELA* constructs like "unless" and "do" [3].

Thus, *Spin* Model Checker can be efficiently used for the formal verification of *BSV* designs. In this work, we show how a *BSV* code of a hardware design can be converted into a *PROMELA* description so that the required properties of the design can then be verified using *Spin* at high-level.

The main advantage of the proposed approach is that it may avoid the application of various abstraction techniques during the verification of hardware designs. Use of various abstraction techniques is very common when RTL models of hardware designs are verified. Thus, the proposed approach aids in faster and efficient verification of *BSV* designs early in the design cycle. In future, we plan to extend this work by formally explaining the translation of *BSV* designs into corresponding *PROMELA* code.

References

[1] Arvind, R. Nikhil, D. Rosenband, and N. Dave, "High-level synthesis: An Essential Ingredient for Designing Complex ASICs," *Proceedings of the International Conference on Computer Aided Design (ICCAD'04)*, pp. 775–782, November 2004.

[2] D. Rosenband and Arvind, "Modular Scheduling of Guarded Atomic Actions," *Proceedings of the Design Automation Conference (DAC'04)*, June 2004.

[3] G. Holzmann, *The SPIN Model Checker*. Addison Wesley, 2004.

[4] E. M. Clarke, O. Grumberg, and D. A. Peled, *Model Checking*. The MIT Press, 2000.

[5] F. Theon and F. Catthoor, *Modeling, Verification and Exploration of Task-Level Concurrency in Real-Time Embedded Systems*. Kluwer Academic Publisher, 2000.

[6] J. C. Hoe and Arvind, "Hardware Synthesis from Term Rewriting Systems," *Proceeding of VLSI'99 Lisbon, Portugal*, December 1999.

Runtime Verification of k-Mutual Exclusion for SoCs

Selma Ikiz
Dept. of Electrical and Computer Engineering
The University of Texas at Austin
Austin, TX 78712, USA
Email: ikiz@ece.utexas.edu

Alper Sen
Formal Verification and Validation Tools,
Design Technology Organization,
Freescale Semiconductor Inc.
7700 W. Parmer Lane, Austin, TX 78729, USA
Email: alper.sen@freescale.com

Abstract— We present an efficient runtime verification environment for detecting mutual exclusion predicates. Such predicates are important for keeping the safe operation of concurrent systems. Our environment models execution traces as partial order traces to increase scalability in runtime verification. We compare two techniques implemented in POTA tool, namely k-exclusion and computation slicing. The k-exclusion problem is a generalization of the mutual exclusion problem in which up to k processes may be in their critical sections at the same time. Our k-exclusion algorithm exploits the fact that if there is a k-exclusion violation then it is impossible to partition events from critical sections into k queues. We earlier presented efficient computation slicing algorithms to detect predicates from a subset of temporal logic CTL. We performed experiments using POTA tool on scalable protocols. Our comparison shows that k-exclusion is substantially better than slicing both in terms of time and space. In all fairness, slicing handles general class of predicates from temporal logic CTL, whereas k-exclusion algorithm handles only a very specific, nonetheless useful, class of mutual exclusion predicates.

I. Introduction

Modern day hardware systems often consist of individual IP blocks connected in meaningful arrangements carrying out independent as well as cooperative objectives. Such Systems-on-Chips (SoCs) are gradually becoming more and more diverse and complicated, thereby gaining popularity. Their enormous potential for re-using IP makes them economically viable solutions to many products out in the market today. The design of a SoC system can be very large and complex. Because SoC's have a large number of components, in general, it is not feasible to give a formal proof of correctness (manual or automatic). Therefore, extensive simulation is still the most common technique used in verifying industrial systems. We advocate the use of *Predicate Detection* (also called Runtime Verification, Assertion Based Verification) that combines formal methods and simulation. Specifically, predicate detection enables *efficient* verification of predicates (or properties) from a formal specification language on execution traces of actual scalable systems.

We also advocate the use of a predicate detection technique that works on *Partial Order Simulation Traces* instead of the traditional total order simulation traces. We use the partial order approach because it has several advantages over the total order approach. Traditional simulation methodologies are woefully inadequate in presence of concurrency and subtle synchronization. The bug in the system may appear only when the delay in synchronization is different from the delay in the simulation trace. A partial order trace model is a more faithful representation of concurrency [10], that is, only the events that have a causal dependency are ordered, e.g. the send of a message and the receive of that message. A partial order encodes possibly exponential number of total orders, therefore we can analyze exponentially more number of traditional traces. Hence, although our predicate detection technique is based on simulation, it becomes scalable in terms of bug detection thanks to partial order traces.

Programming, testing and debugging of concurrent programs is substantially more difficult than that of sequential programs both in terms of correctness and efficiency. Today, almost any large software application is an entire concurrent system made of varying number of processes. These processes often share resources which must be effectively protected against concurrent access. This is usually achieved by executing these actions in critical sections that are protected by semaphores or locks. Hence, an important predicate detection problem is to detect is whether the access to the critical sections are mutually exclusive. This problem is the well known *mutual exclusion* property. To ascertain this safety property, it is vital to detect whether any two processes are in their critical sections simultaneously, i.e., whether there is a mutual exclusion violation in the computation. In this paper, our goal is to detect mutual exclusion predicates based on two algorithms namely k-exclusion and computation slicing. This is the first study that shows some experimental results for k-exclusion algorithm on simulation traces rather than random partial order sets. Moreover, we define bounded sum form (BSF) of predicates and show that BSF can be applied more efficiently than disjunctive normal form (DNF) for some predicates.

The *k-exclusion* problem was posed by Fischer, et al. [5] as a generalization of the mutual exclusion problem in which up to k processes may be in their critical sections at the same time. The k-exclusion property is shown to be a special class of bounded sum predicates [3] and solved by using a max-flow algorithm. Our k-exclusion algorithm exploits the fact that if there is a k-exclusion violation then it is impossible to partition events from critical sections into k queues [4]. We present an efficient, centralized and incremental algorithm.

Computation slicing was introduced in [1], [6], [13] as an abstraction technique for analyzing traces of distributed programs. Intuitively, a *slice* of a trace with respect to a predicate p is a sub-trace that contains all the states of the trace that satisfy p. Note that the set of states that satisfy p may be large, so one could not simply enumerate all the states efficiently either in space or time. A slice contains all the states that satisfy p such that it is computed efficiently (without traversing the state space) and represented concisely (without explicit representation of individual states). Slicing algorithms were developed for temporal logic predicates in [1]. Slicing based predicate detection approach was earlier shown to bring exponential reduction both in terms of time and space for verification of several safety and liveness predicates of scalable protocols [1], [13].

We implemented our k-exclusion algorithm in Partial Order Trace Analyzer (POTA) tool [13], which already implemented slicing based algorithms. We perform experiments on scalable protocols such as Cache Coherence, Distributed Dining Philosophers and PCI based System-on-Chip (SoC). All of these protocols have to satisfy the mutual exclusion property. Our comparison shows that k-exclusion is substantially better than slicing both in terms of time and space. In all fairness, slicing handles general class of predicates from temporal logic CTL, whereas k-exclusion algorithm handles only a very specific, nonetheless useful, class of mutual exclusion predicates.

Related Work: Predicate detection in partial order traces is a hard problem. Detecting even a 2-CNF predicate under EF modality has been shown to be NP-complete, in general [6]. The idea of using temporal logic for analyzing simulation traces has been attracting a lot of attention recently. We first presented a temporal logic framework for partially ordered execution traces in [12] and POTA tool for runtime verification in [13], [14]. Some other examples of using temporal logic for checking execution traces are the MaC tool [9], the Verisim tool [2], and the JMPaX tool [15].

Our work can be applied just as well to concurrent and distributed programs and to behavioral hardware descriptions in Verilog or VHDL. Several EDA companies provide Assertion Based Verification support. However, all those tools use total order simulation trace model, hence the coverage is limited.

II. MODEL

We assume a system consisting of processes denoted by $P_1, \ldots, P_n$. Examples of processes are a thread in a program, node sitting on a PCI bus or a cache in a cache coherence protocol. Processes execute events. Events on the same process are totally ordered. However, events on different processes are only partially ordered. In this paper, we relax the partial order restriction on the set of events and use directed graphs to handle traces and slices with the same model.

We model a *trace* (or a *computation*) as a directed graph, denoted by $\langle E, \rightarrow \rangle$, with vertices as the set of events E and edges as $\rightarrow$. We use event and vertex interchangeably. To limit our attention to only those consistent global states that can actually occur during an execution, we assume that the paths in $\langle E, \rightarrow \rangle$ contains at least the partial order relation.

A partial order relation known as Lamport's happened-before relation [10] has been used for modeling simulation traces. We use a mechanism known as *vector clocks* to represent the partial order relation. A vector clock assigns timestamps to events such that the partial order relation between events can be determined by using the timestamps.

III. K-EXCLUSION ALGORITHM

The k-exclusion problem was posed by Fischer, et al. [5] as a generalization of the mutual exclusion problem in which up to k processes may be in their critical sections at the same time. The k-exclusion property is shown to be a special case of bounded sum predicates [3] and solved by using a max-flow algorithm. Here we define bounded sum form (BSF) of predicates to compare them in disjunctive normal form (DNF). Some examples of predicates where BSF can be applied more efficiently than DNF:

1) Two process mutual exclusion:
   (DNF) $\bigvee_{i,j \in 0\ldots(n-1)} (\mathsf{EF}(CS_i \wedge CS_j))$
   (BSF) $(\mathsf{EF}(CS_0 + CS_1 + \ldots CS_{n-1} > 1))$
2) At least one server is available:
   (DNF) $(\mathsf{EF}(\neg avail_1 \wedge \neg avail_2 \wedge \ldots \neg avail_n))$
   (BSF) $(\mathsf{EF}(\neg avail_1 + \neg avail_2 + \ldots \neg avail_n > n-1))$
3) K process mutual exclusion:
   (DNF) $\bigvee_{i_0, i_1, \ldots, i_K \in 0\ldots(n-1)} (\mathsf{EF}(CS_{i_0} \wedge CS_{i_1} \wedge \ldots CS_{i_K}))$
   (BSF) $(\mathsf{EF}(CS_0 + CS_1 + \ldots CS_{n-1} > K))$

The k-exclusion violation can be detected by checking whether more than k processes have executed their critical section simultaneously (concurrently). However, a detection algorithm that uses DNF form requires $C(n, k+1)$ combination checks on computation trace for each disjunct. On the other hand using a max-flow algorithm has $O(E^2 . log(E))$ complexity. Our algorithm described below has a better complexity.

Our k-exclusion algorithm exploits the fact that if there is a k-exclusion violation then it is impossible to partition critical section events into k queues (Dilworth's Chain Partition) [4]. The k-exclusion algorithm is centralized and incremental. We call the process *checker process* if it is executing the algorithm (central process). Each process, whenever changes its state, sends its local state along with its vector clock to the checker process. When a new event arrives, the checker process executes the algorithm if the new event is a critical section event, otherwise discards it. The k-exclusion algorithm rearranges the queues according to the new event and checks whether the number of queues in the resultant arrangement is more then k or not. The algorithm assumes that when a new event arrives, all events that happened-before it have already arrived and been processed. This is achieved by buffering the new event if it violates the assumption and processing it later. Whether to buffer an event can be determined efficiently by examining its vector clock.

Setup: The algorithm uses queues to store the events. Events are stored in an increasing order so that head is the smallest event and tail is the largest event in the queue. It keeps two type of queue compositions: work and history. We refer to the set of work and history queues as work and history space, respectively. W^j represents the work space when jth element arrives, while H^j represents the history space.

Execution: Whenever a new event arrives, k-exclusion tries to append it to one of the queue tails. However, this might not be always possible since new event may be concurrent to all the queue tails. In this case, it calls the $Merge$ function to merge s queues into $s-1$, if possible. The $Merge$ function takes s queues as input and returns three type of queue sets; input queues Q, output queues O, and history queues H'. K-exclusion updates its work space according to input queues of $Merge$, if there are s events that are mutually concurrent. Otherwise, it uses the output queues of $Merge$. The history space is updated by the history queues of $Merge$. Note that $s \leq k+1$ at all times.

Merge Function: The $Merge$ function begins with s queues called *input* queues and $s-1$ empty queues called *output* queues, and a spanning tree that is formed by the input and the output queues. The spanning tree has exactly k vertices and $k-1$ edges. An edge corresponds to an output queue and a vertex corresponds to an input queue. Therefore each edge has a label which identifies the output queue it corresponds with. No labels are duplicated in the tree, thus each output queue is represented once. Similarly, each input queue is represented exactly once. At each step, it removes some events from the input queues and append them to the output queues. This relocating procedure continues until $Merge$ finds s concurrent events or one of the input queues becomes empty. The spanning tree is used in the decision process of which output queue to append the events. The algorithm maintains the following invariant: If an edge between vertices q_i and q_j is labeled as o_k, then the heads of q_i and q_j are bigger than the tail of o_k. Intuitively, at a merge step, queue head a_i is selected for removal if it is less than one of the other input queue heads. $Merge$ updates its history and output queues if it witnesses s or $s-1$ concurrent events. It moves all the elements in output queues to history queues. Splitting the events into work and history queues reduces the number of comparisons for later steps. After each $Merge$ call, the heads of queues in the work space are mutually incomparable. Any event happened before them are placed into the history space. This splitting does not affect the future comparisons and we prove that it is sufficient to compare the new event to the events in work space [8].

Merge Example: An example for $Merge$ is given in Figure 1 to show how the state of the tree and of the queues are modified at each step. Figure 1a shows the initial setup. Q, O, and H' show the input, output and history queues, respectively, while dashed lines show the edges in the spanning tree. Remember that an edge labeled as an output queue between input queues states that the tail of the output queue is less than both of the input queue heads.

Step 1: [1,0,0] is less then [2,0,0], and [1,1,0] and [2,0,0] are

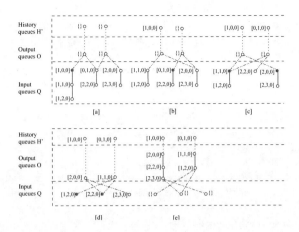

Fig. 1. An example for Merge

concurrent. Hence, [1,0,0] is placed in history queues (see Figure 1b).

Step 2: [0,1,0] is less then [1,1,0], and [1,1,0] and [2,0,0] are concurrent. Hence, [0,1,0] is placed in history queues (see Figure 1c).

Step 3: [2,0,0] and [1,1,0] are both less then [2,2,0]. They are placed in output queues (see Figure 1d).

Step 4: [1,2,0] and [2,2,0] are both less then [2,3,0]. They are placed in output queues. Since there is an empty input queue, $Merge$ places the remaining events adhering to the spanning tree (see Figure 1d).

k-exclusion example: Consider the computation in Figure 2. Assume that the order of events arrived to checker process is $x_2, x_1, x_3, x_4, x_5, x_6, x_7, x_8, x_9, x_{10}$. We demonstrate two steps where splitting the queues happens since other steps are trivial.

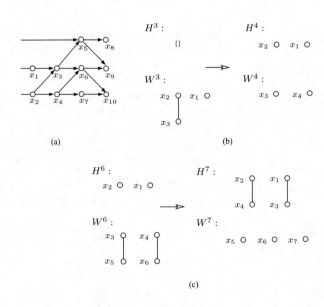

Fig. 2. (a) A computation (b) transition in steps 3 & 4 (c) transition in steps 6 & 7

1) Merge is successful: At time 4, the checker has seen $X^3 = \{x_2, x_1, x_3\}$ and the new event is x_4. There are two queues in the work space and its content is as follows: q_1 contains x_2 and x_3, and q_2 contains x_1. There is no queue to append x_4 since both x_1 and x_3 are incomparable with x_4, hence k-exclusion creates a new queue q_3 and appends x_4 to q_3, and invokes $Merge$. Observe that s is three. $Merge$ places x_2 in o_1 and x_1 in o_2, and observes x_3 and x_4 as concurrent. Since $s-1$ is two, it appends the contents of output queues to history queues. It places x_3 in o_1, x_4 in o_2, and returns queues back, input queue being empty. Then, k-exclusion assigns the output queues as work queues, and appends the history queues of $Merge$ to its history space. Figure 2b shows the transition between time 3 and 4.

2) There are s mutually concurrent events: At time 6, the checker has seen $X^6 = \{x_2, x_1, x_3, x_4, x_5, x_6\}$ and the new event is x_7. There are two queues in the work space and their contents are as follows: q_1 contains x_3 and x_5, and q_2 contains x_4 and x_6. There is no queue to append x_7 since both x_5 and x_6 are incomparable with x_7, hence k-exclusion creates a new queue q_3 and appends x_7 to q_3, and makes a $Merge$ call. $Merge$ places x_3 in o_1 and x_4 in o_2, and computes x_5, x_6, and x_7 as concurrent. Since s is three and there are three mutually concurrent events, no reduction is possible. $Merge$ appends the contents of output queues to history queues, and returns back. Then, k-exclusion assigns the input queues as work queues, and appends the history queues of $Merge$ to its history space. Figure 2c shows the transition between time 6 and 7.

The worst case complexity of k-exclusion algorithm is $O(k|E|^2)$, where E is the set of critical section events, and k is the allowed number of concurrent events. When k is one, the complexity reduces to $O(|p|.|E|)$, where $|p|$ is the number of temporal operators in a bounded sum predicate p in BSF. The correctness discussion of the algorithm can be found in [8].

IV. COMPUTATION SLICE

Informally speaking, a computation slice (or a slice) is a concise representation of all those global states of the computation that satisfy a global predicate (or simply predicate or property).

In the next two sections, we will illustrate the basic slicing algorithm on a simple example. A formal treatment of slicing algorithms can be found in [1].

A. Non-temporal Predicate Slicing

Figure 3c shows the slice of the computation in Figure 3a with respect to the non-temporal predicate $CS_1 \wedge CS_2$, which is a conjunction of two local predicates from the processes P_1 and P_2, respectively. We say that a predicate is *local* if it only depends on variables of a single process. In the figure, if a local predicate is true for an event of a process, we represent it using an empty circle, otherwise with a filled circle. Intuitively, to obtain the slice with respect to a local predicate, we add an edge from the successor of a filled circle back to it, thereby increasing the number of incoming

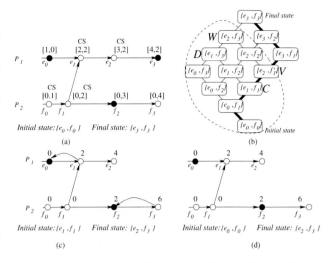

Fig. 3. (a) A computation (b) its set of all reachable global states (c) its slice with respect to $CS_1 \wedge CS_2$ (d) its slice with respect to $\mathsf{EF}(CS_1 \wedge CS_2)$

neighbors and disabling the local event that does not satisfy the predicate from being a consistent global state of the newly obtained trace. Now, we will further demonstrate why these additional edges give us the slice by showing that addition of edges results in removing consistent global states of the computation that do not satisfy the predicate. Observe that, the following global states, $\{e_0, f_0\}, \{e_0, f_1\}, \{e_0, f_2\}$, and $\{e_0, f_3\}$ of the computation in Figure 3c depicted in Figure 3b, do not satisfy the local predicate CS_1. This is because in all of these four states P_1 has executed only its initial event, where local predicate CS_1 is false. We then add the edge (e_1, e_0) in Figure 3c. From the definition of a consistent global state, any consistent global state of the slice, upon including e_0 must include e_1, which is an incoming neighbor. Since all of the global states $\{e_0, e_1\}, \{e_0, f_1\}, \{e_0, f_2\}, \{e_0, f_3\}$ include e_0, they must have included e_1 as well, which they do not. Hence, they cannot belong to the consistent global states of the slice. Also, note that if we added an edge from (e_2, e_0) instead of (e_1, e_0), we would have eliminated states such as $\{e_1, f_1\}$, which satisfy the predicate, from the slice. This is because any consistent global state upon including e_1 must include e_2, due to transitivity of the incoming neighbors. Similarly, we can reason about the additional edge (f_3, f_2) in the slice. As a result, from the thirteen consistent global states in Figure 3b, by adding two edges and not traversing the state space, this exercise eliminates nine – retaining states $\{e_1, f_1\}$, $\{e_1, f_3\}, \{e_2, f_1\}$, and $\{e_2, f_3\}$, which are the only states of the computation that satisfy $CS_1 \wedge CS_2$. □

It has been shown that the slice exists for all predicates, however it is, in general, intractable to compute the slice for an arbitrary predicate [6]. Our approach to computation slicing is based on exploiting the structure of the predicate itself.

B. Temporal Predicate Slicing

To illustrate predicate detection using computation slicing, consider the computation in Figure 3a again. Let $p = CS_1 \wedge CS_2$, and suppose we want to detect $\mathsf{EF}(p)$, that is, whether there exists a global state of the computation that satisfies p. Without computation slicing, we are forced to examine all global states of the computation, thirteen in total, to decide whether the computation satisfies the predicate. Alternatively, we can compute the slice of the computation with respect to the predicate $\mathsf{EF}(p)$ and use this slice for predicate detection. For this purpose, first, we compute the slice with respect to p as explained above. The slice is shown in Figure 3c. The slice contains only four states C, D, V and W (see Figure 3b) and has much fewer states than the computation itself – exponentially smaller in many cases – resulting in substantial savings. Next, using the slice in Figure 3c, we can obtain the largest reachable state that satisfies p in the computation, which is denoted by W. We also know from the definition of $\mathsf{EF}(p)$ that every global state of the computation that can reach W satisfies $\mathsf{EF}(p)$, e.g., states enclosed in the dashed ellipse in Figure 3b. Therefore, applying this observation we can compute the slice with respect to $\mathsf{EF}(p)$ as shown in Figure 3d. In the slice, there are less number of vertices than the computation and the slice and the computation have the same set of global states up-to W. Finally, we check whether the initial state of the computation is the same as the initial state of the slice, since predicate detection is concerned whether the initial state of a computation satisfies a given predicate. If the answer is yes, then the predicate is satisfied, otherwise not, and a counter-example is returned. In this case the predicate p is satisfied, that is mutual exclusion is violated.

Note that the above temporal slicing algorithm returns all consistent global states of the computation that satisfies the given predicate not just one.

V. PREDICATE DETECTION USING SLICING

We present efficient computation slicing algorithms for a subset of temporal logic CTL in [1]. The slices are computed recursively starting from the deepest nesting of predicates and applying the appropriate slicing algorithm, either temporal or boolean, while proceeding outwards.

Predicate detection is concerned about checking whether the initial consistent cut of a computation satisfies a given predicate. Since the slice contains all consistent cuts of the computation that satisfies the predicate, upon computing the slice, we simply check whether the initial consistent cut of the computation belongs to the slice. This can be done by ascertaining whether both initial cuts of the computation and the slice are the same.

Our predicate detection algorithm for a k-exclusion predicate in DNF is as follows. First, we compute slices wrt each conjunctive predicate p in every disjunct. Second, we obtain the slices wrt EF for every disjunct. Third, we check whether the initial cut of the computation is the same as the initial cut of the slice wrt $\mathsf{EF}(p)$ for any disjunct.

The complexity of our trace based predicate detection technique is polynomial in n, that is, $O(|p| \cdot n^2 |E|)$, where $|p|$ is the number of boolean and temporal operators in a predicate p, E is the set of vertices and n is the number of processes. However, this complexity is for a general class of temporal logic predicates. When p is a 1-exclusion predicate, the complexity can be reduced to $O(n|E|)$, since the complexity of slicing wrt conjunctive predicates is $O(|E|/n)$ and instead of computing the slices wrt $\mathsf{EF}(p)$, we check if the slice for p is empty or not. Similarly, the complexity is $C(n, k+1).(k+1).|E|/n$ for k-exclusion.

Note that slicing leads to efficient algorithms even for predicates such as $p_1 \wedge p_2$, where p_1 has efficient slicing algorithms but p_2 does not such as $p_2 = x_1 + x_2 * x_3 < 7$. This is because it is better to detect p_2 on the slice for p_1 instead of the computation, using say an exhaustive search algorithm, since the the state space of the slice is much smaller than the state space of the original computation.

VI. EXPERIMENTS

We ran experiments with our tool Partial Order Trace Analyzer (POTA) with general slicing algorithms and the k-exclusion algorithm for monitoring execution traces of programs for mutual exclusion violations. The tool contains three modules; analyzer, translator, and instrumentor. POTA instruments the given program by inserting code at the appropriate places in the description to be monitored. The instrumented description is such that it outputs the values of variables relevant to the predicate in question and keeps a vector clock. POTA also implements computation slicing and k-exclusion algorithms. All experiments were performed on a 3.06 Ghz Xeon processor machine running Linux. We restricted the memory usage to 512MB, but did not set a time limit. The two performance metrics we measured are running time and memory usage. In the case of slicing both metrics also include the overhead of computing the slice. Further experimental results can be obtained from the website [11]. As experimental testbeds, we chose distributed dining philosophers, cache coherence and PCI protocols. We could stretch our verification algorithms by incrementing both the complexity and the number of processes in the systems, which also demonstrates the effectiveness of the algorithms for very large state spaces and complex designs.

A. Distributed Dining Philosophers

We use the Java protocol from [7] for this exercise. The protocol consists of multiple philosophers who sit around a table and spend their time thinking and eating. However, a philosopher requires shared resources, such as forks, to eat. The protocol coordinates access to the shared resources. Each philosopher has 3 local states namely *think*, *hungry*, and *eat*. The philosophers do not have a central server that they can query for fork availability. Instead each philosopher has a servant who communicates with the two neighboring servants to negotiate the use of the forks. The servants send "need left fork", "need right fork", "pass left fork", and "pass right fork" messages back and forth. Each fork is always in the possession of some philosopher, one of the two on either side of the fork. When a philosopher finishes eating, it labels its two forks as dirty. A hungry philosopher's servant is required to give up a dirty fork in its possession, if asked for by its hungry neighbor's servant. This prevents starvation. We check the following property. We require mutually exclusive use of forks, that is, a shared resource should not be used by more than one philosopher at a time. This can be ascertained by checking whether two neighbor philosophers are eating at the same time. This safety property can be stated as $\bigwedge_{i,j \in 0...(n-1)} (\mathsf{AG}(\neg eat_i \vee \neg eat_j))$, where eat_i denotes that philosopher i is in eating state and j denotes the neighbor of philosopher i. We can check whether this property is violated by checking the complement of the safety property, which is $\bigvee_{i,j \in 0...(n-1)} (\mathsf{EF}(eat_i \wedge eat_j))$.

Figure 4 displays our results for this property.

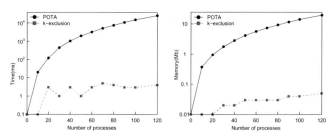

Fig. 4. Dining philosophers verification results for mutual exclusion property

B. Cache Coherence Protocol

The MSI (Modified Shared Invalid) cache coherence protocol is a protocol to maintain data consistency among a number of caches connected to a central directory structure in a multiprocessor system. The protocol is a directory based scheme in which individual processes snoop on all other processors' activities over a shared directory. The details of the protocol can be found in [11].

The property we checked on the MSI protocol is the safety property, "two caches cannot be in the modified state simultaneously". The complement of the property is $\mathsf{EF} smodified_i \wedge modified_j$, where i and j are cache identifiers.

Figure 5 displays our results for this property. k-exclusion took two millisecond and 1 MB to complete for 120 processes, whereas, POTA took 390 seconds and 220 MB. Due to the overhead associated in generating traces, we stopped generating traces for more than 120 processes.

C. PCI

The PCI Local Bus is a high performance bus with multiplexed address and data lines. It is used as an interconnect mechanism between peripheral controller components or add-in boards and processor/memory systems. Our example contains an arbiter, and a parameterizable number of nodes that can act either as a master or a target machine.

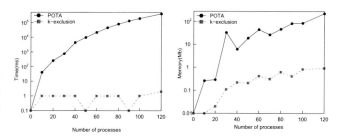

Fig. 5. MSI verification results for for mutual exclusion property

In the SoC based on the PCI backbone, we have a parameterizable number of devices that are non-deterministically sending requests to the arbiter for accessing the bus as a master and interacting with each other. The safety property that we verified is "there cannot be two masters of the bus". Symbolically, $\mathsf{EF}(master_i \wedge master_j)$, where i and j are device identifiers.

For this experiment, we have varied the number of nodes from 10 to 120 for the safety property The reason for the less number of nodes in the safety property experiments is that the property, given in DNF, size is larger in this case (that is, every possible combination of nodes i and j exists in the property). The related time and memory usage is displayed in Figure 6 for safety property. k-exclusion took two millisecond and 3 MB to complete for the worst case, whereas, POTA took 640 seconds and 390 MB.

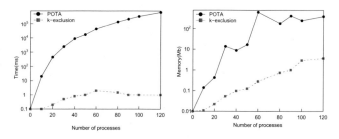

Fig. 6. SoC verification results for mutual exclusion property

VII. CONCLUSION

We presented a comparison of two techniques for detecting mutual exclusion violations. Our experimental results using POTA tool show that the the specialized k-exclusion algorithm performs substantially better than the slicing based algorithm. In all fairness, the slicing based algorithms are meant for general classes of predicates, hence do not exploit the properties of mutual exclusion predicates as k-exclusion algorithm does. Also, although the complexity of the slicing based algorithm specifically for 1-exclusion is similar to that of k-exclusion algorithm, we believe that due to the implementation overhead we obtain poor slicing performance. Nonetheless, for mutual exclusion predicates we have a very efficient and scalable algorithm implemented in POTA. For future work, we would like to investigate whether the ideas used in k-exclusion algorithm can be used for other types of predicate detection as well.

References

[1] A. Sen and V. K. Garg. Detecting Temporal Logic Predicates in Distributed Programs Using Computation Slicing. In *Proceedings of the 7th International Conference on Principles of Distributed Systems (OPODIS)*, La Martinique, France, December 2003.

[2] K. Bhargavan, C. A. Gunter, I. Lee, O. Sokolsky, M. Kim, D. Obradovic, and M. Viswanathan. Verisim: Formal Analysis of Network Simulations. *IEEE Transactions on Software Engineering*, 28(2):129–145, 2002.

[3] C. Chase and V. K. Garg. Efficient Detection of Restricted Classes of Global Predicates. In *Proceedings of the Workshop on Distributed Algorithms (WDAG)*, pages 303–317, France, September 1995.

[4] B. A. Davey and H. A. Priestley. *Introduction to Lattices and Order*. Cambridge University Press, Cambridge, UK, 1990.

[5] M. Fischer, N. Lynch, J. Burns, and A. Borodin. Resource Allocation with Immunity to Process Failure. In *Proceedings of the 20th IEEE Symposium on the Foundations of Computer Science (FOCS)*, pages 234–254. IEEE Computer Society Press, 1979.

[6] V. K. Garg. *Elements of Distributed Computing*. John Wiley & Sons, 2002.

[7] S. Hartley. *Concurrent Programming: The Java Programming Language*. Oxford University Press, 1998.

[8] S. Ikiz and V. K. Garg. Efficient Incremental Optimal Chain Partition of Distributed Program Traces. In *Proceedings of the IEEE International Conference on Distributed Computing Systems (ICDCS)*, Lisbon,Portugal, July 2006.

[9] M. Kim, S. Kannan, I. Lee, O. Sokolsky, and M. Viswanathan. Java-MaC: a Run-time Assurance Tool for Java Programs. In *Proceedings of the 1st International Workshop on Runtime Verification (RV)*, volume 55 of *ENTCS*. Elsevier Science Publishers, 2001.

[10] L. Lamport. Time, Clocks, and the Ordering of Events in a Distributed System. *Communications of the ACM (CACM)*, 21(7):558–565, July 1978.

[11] Partial Order Trace Analyzer (POTA) web site, 2003. http://maple.ece.utexas.edu/~sen/POTA.html.

[12] A. Sen and V. K. Garg. Detecting Temporal Logic Predicates on the Happened-Before Model. In *Proceedings of the 16th International Parallel and Distributed Processing Symposium (IPDPS)*, Fort Lauderdale, Florida, April 2002.

[13] A. Sen and V. K. Garg. Partial Order Trace Analyzer (POTA) for Distributed Programs. In *Proceedings of the 3rd International Workshop on Runtime Verification (RV)*, volume 89 of *Electronic Notes in Theoretical Computer Science*. Elsevier Science Publishers, 2003.

[14] A. Sen and V. K. Garg. Formal Verification of Simulation Traces Using Computation Slicing. *IEEE Transactions on Computers*, 56(4), 2007.

[15] K. Sen, G. Rosu, and G. Agha. Runtime Safety Analysis of Multithreaded Programs. In *Proceedings of the Symposium on the Foundations of Software Engineering (FSE)*, 2003.

A Scalable Symbolic Simulator for Verilog RTL

Sasidhar Sunkari[*], Supratik Chakraborty[†], Vivekananda Vedula[‡] and Kailasnath Maneparambil[§]

[*] Intel, Bangalore
Email : sasidhar.sunkari@intel.com
phone: +91 80 26053655

[†] Indian Institute of Technology, Bombay
Email : supratik@cse.iitb.ac.in

[‡] Intel, Bangalore
Email : vivekananda.vedula@intel.com

[§] Intel, Arizona
Email : kailasnath.s.maneparambil@intel.com

Abstract

Symbolic simulation is an important technique used in formal property verification and test generation for digital circuits. Existing symbolic simulators predominantly operate at the gate level, reasoning about individual bits and signals. As a result, their performance does not scale well to large circuits like microprocessors. Word-level symbolic simulators address this problem to some extent, but present other challenges, such as fixpoint detection when simulating multiple modules that mutually trigger each other. In this paper, we present some exploratory ideas for performing word-level symbolic simulation over a Verilog RTL description of a circuit. We outline the basic technique of simulation and of handling fixpoints, discuss issues faced in our approach and present solution techniques to counter these issues. We also present initial experimental results obtained by applying our algorithms to a Verilog model of an x86 processor design.

I. Introduction

Simulation with specific test inputs is the predominant technique used in functional verification of digital circuits. Unfortunately, it is impractical to exhaustively simulate every possible execution of large and complex circuits like microprocessors. Symbolic simulation offers a promising alternative by combining the flexibility and scalability of conventional simulation with the analytical power of sophisticated symbolic methods. In symbolic simulation, we replace multiple simulation runs, each with different inputs, with a single run in which the inputs are assigned symbolic values. The output of a run of symbolic simulation is a set of symbolic expressions representing possible values of circuit nodes at different time instants. By reasoning about these symbolic expressions, it is often possible to infer properties of the circuit that would have otherwise required a large number of specific input based simulation runs to discover. Symbolic simulation has been used in earlier work for formal and semi-formal property verification, as well as for test case generation. Early symbolic simulators that were successfully applied on non-trivial circuit designs dealt with simulation at the switch or gate-level [1], [2], [3], [4], and reasoned about individual bits or signals. For circuits of the scale of microprocessors, however, reasoning at the level of switches and individual bits is prohibitively expensive. It is therefore necessary to simulate such large circuits at the word-level, starting from a Register Transfer Level (RTL) description of the design. While there has been earlier work on word-level symbolic simulation for RTL designs [5], these approaches have primarily restricted the simulation to a pre-determined (or

This work was supported by Intel Corporation, and was done while Sasidhar Sunkari was at Indian Institute of Technology, Bombay.

user-provided) number of simulation cycles. This works well if all interesting signal transitions at all circuit nodes happen within the pre-determined number of cycles. Several RTL designs, however, have a set of blocks that mutually trigger each other. In such cases, it may be difficult to statically determine how many simulation cycles must elapse before the values of all nodes stabilize. Hence, fixing the number of simulation cycles a priori may not be the right way to symbolically simulate such designs. In this paper, we describe our ongoing effort towards building a symbolic simulator for Verilog RTL designs that addresses the above issue through a fixed point formulation, and also scales well in performance with the size of the RTL description. This work is part of a larger project that aims to use symbolic simulation and approximate constraint solving techniques for test instruction generation for microprocessor designs. However, the focus of this paper is on the design of the symbolic simulator.

To illustrate the potential advantages of word-level symbolic simulation over bit-level simulation, consider the example shown in Fig. I. This example has an adder with inputs I_1 and I_2. When the circuit is started, the values of I_1 and I_2 are provided by the user through the multiplexers at the inputs of the adder. The arithmetic sum obtained at the adder output is loaded into register R at the next rising edge of the clock. The output of the register is then fed to a left shifter (a combinational circuit) T, that effectively multiplies its input by 2. The values of I_1 and I_2 in subsequent cycles of operation are obtained from the outputs of R and T respectively through the multiplexers. We assume that all operands are treated as unsigned integers, all datapaths are 64 bits wide, and when overflow occurs, the least significant 64 bits are retained.

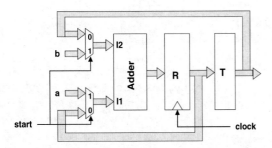

Fig. 1. Block diagram of a sequential circuit

Suppose the initial symbolic values of I_1 and I_2, as given by the user, are a and b respectively. *Assuming no overflows, i.e. all symbolic expressions evaluate to values between $2^{64} - 1$ and 0,* the word-level simulation of the above circuit for the first four clock cycles is shown in Table I. For clarity of presentation, the expressions in the Table have been simplified arithmetically. These expressions can now be used for performing various analyses of the circuit in Fig. I. Specifically, a property that is inferred from these expressions holds true for all simulation runs starting from values of a and b that don't result in overflows in the first four cycles. Thus the result of symbolic simulation can be used for property checking. The expressions obtained above may also be used to generate test cases or to prove untestability of specific test goals. For example, suppose we wish to find values of inputs a and b such that the most significant bit (MSB) of register R is 1 in the fourth clock cycle without any overflows in any datapath operation. From Table I we find that the expression for R in the fourth clock cycle is $27(a+b)$. Since register R is 64-bits wide, we require $27(a+b) \geq 2^{63}$. To ensure that the output of the adder doesn't overflow in the fourth clock cycle, we also require $81(a+b) \leq 2^{64} - 1$, so . Unfortunately, both these constraints cannot be simultaneously satisfied. Therefore it is impossible to obtain test inputs that will set the MSB of R to 1 in the fourth clock cycle without causing any overflows. In fact, since the output of the adder in any cycle (≥ 1) is three times the value of R in the same cycle, it is impossible to have the MSB of R set to 1 without the output of the adder overflowing, if all data buses have the same width.

The above example illustrates how word-level symbolic simulation scores over bit-level symbolic simulation. If we were to solve the same test generation problem by symbolically simulating the circuit at the level of individual bits, we would have generated a separate expression for each bit at the output of each block. These expressions would typically be represented as Binary Decision Diagrams (BDD) or propositional formulae in Conjunctive Normal Form (CNF). Next, a SAT solver or BDD-based engine would be used to check the propositional satisfiability of the expression generated for the most significant bit of register R. In general, this may involve far more computational effort than the above analysis based on word-level symbolic expressions, in which we reasoned using properties of bit-vectors or integers instead of using bit-level reasoning. Note also that the word-level expressions indicate the impossibilty of finding a test input that sets the MSB of R to 1 without causing any overflows, independent of the number of clock cycles and width of the datapath, as long as all data buses have the same width. It would be very difficult to arrive at the same conclusion using bit-level expressions. This demonstrates the power of symbolic reasoning at the word-level.

The focus of this paper is on the design and development of a word-level symbolic simulator for Verilog RTL designs. Our simulator takes as input a circuit design in Verilog and performs word-level simulation of the various behavioral constructs. This generates symbolic expressions encoding possible values of different circuit nodes. The

TABLE I. Symbolic simulation of circuit in Fig. I

Clock	I_1	I_2	Adder output	R	T
0	a	b	$a+b$	Unspecified	Unspecified
1	$a+b$	$2(a+b)$	$3(a+b)$	$(a+b)$	$2(a+b)$
2	$3(a+b)$	$6(a+b)$	$9(a+b)$	$3(a+b)$	$6(a+b)$
3	$9(a+b)$	$18(a+b)$	$27(a+b)$	$9(a+b)$	$18(a+b)$
4	$27(a+b)$	$54(a+b)$	$81(a+b)$	$27(a+b)$	$54(a+b)$

following sections describe the various issues faced in developing our prototype simulator, and discusses techniques used to overcome these issues. Section II outlines related work in symbolic simulation for property verification and test case generation for digital circuits. Our approach to performing word-level symbolic simulation is then presented in section III. An interesting sub-problem that needs to be addressed is the identification of sets of `always` blocks that mutually (or recursively) trigger each other. Section IV describes an efficient and approximate technique used in our simulator for identifying such `always` blocks. This allows our simulator to scale to large designs while conservatively identifying `always` blocks that are recursively triggered. Section V presents some initial experimental results on a Verilog x86 design. Finally, Section VI concludes the paper with a note on directions for future work.

II. Related Work

Formal verification of switch and gate-level digital circuits using symbolic simulation and a ternary valued lattice was extensively studied by Bryant and Seger in their seminal work on Symbolic Trajectory Evaluation(STE)[1], [2], [3]. This technique has subsequently been implemented in tools like COSMOS, and has been successfully applied to verify the correctness of several circuits [4]. In ternary valued symbolic simulation, expressions at circuit nodes are often encoded and represented using BDDs. Wilson et al [6] studied the use of approximations to reduce the sizes of BDDs in bit-level symbolic simulation. Their approach selectively assigns approximate values (combinations of 0, 1 and X) to internal nodes in a circuit, when such approximate assignments do not affect the functionality being tested or verified.

Su et al [7] used a form of symbolic simulation and special logic transformation techniques to automatically identify invariants in finite-state transition systems. Zeng et al used symbolic simulation and word-level satisfiability checking to generate functional test vectors [5]. They considered using both an integer linear program formulation and a constraint logic program formulation for word-level satisfiability checking. In general, solving a set of constraints on expressions obtained from symbolic simulation involves reasoning about multiple theories like those of bit-vectors, arrays, bounded integers, uninterpreted functions, etc. The development of powerful Satisfiability-Modulo-Theories (SMT) solvers in recent years has greatly aided the use of symbolic simulation and constraint solving based approaches for property verification and test generation of digital circuits..

The theory of abstract interpretation, first formulated by Cousot and Cousot [8], has been used with significant success to tame the practical complexity of verification in the context of sequential programs. In this approach, a concrete domain of values is mapped to a "simpler" abstract domain that permits easier reasoning, albeit at the cost of some precision loss. Once the analysis has been performed in the abstract domain, the results can be re-interpretted in the concrete domain, often helping in proving properties of complex systems. The theory of abstract interpretation guarantees the soundness of any such verification result, although there are no completeness guarantees. Since word-level symbolic simulation may also be looked upon as abstract execution of a program in a hardware description language (HDL), the theory of abstract interpretation offers promising ways to deal with the complexity of word-level symbolic simulation.

Despite some industrial interest, symbolic simulation is however yet to see widespread usage in commercial tools. Among the commercial offerings that use symbolic simulation technology, Blue Pearl Software has a tool for checking whether a chip meets its timing goals. This tool uses proprietary technologies for RTL analysis, high level symbolic simulation and state space exploration. Innologic (currently part of Synopsys) uses symbolic simulation for checking equivalence of designs at different levels of description. Nusym is yet another company that makes use of symbolic simulation and analysis techniques for automation of test extraction, coverage estimation and coverage convergence.

III. Word Level Symbolic Simulation

The symbolic simulator described in this paper simulates only the synthesizable subset of Verilog. Synthesizable Verilog has certain restrictions which make the design of the simulator easier. For example, the number of iterations for each loop statement is known at compile time and hence they can be unrolled before the expressions are generated . The behavioral code thus contains only the conditional statements and assignments.

The symbolic expressions are generated by parsing all the control flow paths in each behavioral block (always statement). The input data structure used for word-level simulation is explained in III-A.

A. Control Data Flowgraph

In word-level symbolic simulation, the circuit model is a description of the circuit at RTL. The given RTL design is translated into a collection of interconnected *Control Data Flow Graphs*(CDFGs). Each CDFG represents the control and data flow within an always block of a module. Thus we would have as many CDFGs as the number of always blocks in the design. A representative node for each module contains the rest of information like continuous assignments, links to CDFGs of all always blocks within the module, etc. The representative nodes for each module are connected by instance nodes according to the instantiations made in the design. The port connectivity information is contained in these instance nodes. The expressions are generated by traversing the CDFG and executing the statements in accordance with the semantics of Verilog language[9]. 'Execution of a statement' here means symbolic manipulation of words. The generated expressions have word level operators like arithmetic add, left shift, etc which are used in the RTL design. We define a *word* as *"a contiguous vector of bits in a variable that remains undivided through out the design"*. Alternately, a *word* is *"a bit vector whose bits are read or written all at once"*, that is, there does not exist a case where only a part of the word is read/written and the remaining are not. For example, considering a and b to be two 32-bit variables, the following lines of Verilog code
a[0:9] = 10b'1;
a[10:31] = b[10:31];
causes the division of words as a[0:9], a[10:31], b[0:9] and b[10:31].
Also, if we have an assignment
x[0:9] = y[0:9]
where x[0:9] is unfragmented but y[0:9] is fragmented into y[0:4] and y[5:9], then we also breakdown x[0:9] into x[0:4] and x[5:9] and make the assignments
x[0:4] = y[0:4] and x[5:9] = y[5:9]

The Language Reference Manual for Verilog [9] describes a simulation algorithm which is based on discrete event model and uses a stratified queue. The value updates to variables in the design causes events to be generated which are enqueued in a priority queue. Word-level symbolic simulation was tried using this algorithm to simulate an example design. Informally, the simulation algorithm is to traverse all the control paths of the top module and execute the respective statements(continuous assignments, port assignments and always blocks) which are triggered due to assignments made in the traversal. However it was observed that *the simulation never terminated!* This section describes the reason for the failure of the naive simulation algorithm. It presents the challenges posed by the *always* statement in performing symbolic simulation and presents our approach of dealing with them.

The major issue that was encountered was to detect a change in a symbolic expression which is present in the sensitivity list of an always statement. In other words, given two word level symbolic expressions E_1 and E_2, we wish to say whether these two expressions are equivalent or not. In our implementation, it was assumed that an assignment to a variable, would have possibly changed the value of the variable and executed statements that are sensitive to the variable being assigned. However, there could exist synthesizable Verilog designs where an always block execution can possibly cause a series a excitations which propagate back to the sensitivity list of the always block itself and thus re-trigger itself. Such a case leads to an infinite sequence of executions and hence the simulation algorithm never terminated. The below example is a case where such an infinite triggering is possible.

module $A(a,b,c)$	**module** $B(x,y)$
input[0:2] b,c;	input[0:2] x;
output reg[0:2] a;	output reg[0:2] y
reg[0:2] x,y;	**always**@(x)
B b(x,y);	**if**(x==3b'100)
always@(b or c or y)	y=x<<1
if(b==3b'000)	**endmodule**
x = c << 1	
else	
a = y & b	
endmodule	

Let us assume that we are symbolically executing the always block in module A which causes an assignment to the variable x. This value change propagates to the instance of module B and causes its always block to execute. This execution assigns a value to the variable y which is propagated back to the parent module A. Now, since the always block in module A is sensitive on y, it is executed again. Thus, the chain of events can repeat for an infinite number of times.

It is possible to define an **Excitation Dependency Graph** to depict the chain of events that can possibly cause an always block to recursively trigger itself. The nodes of an excitation dependency graph are the words of each instance and the always blocks of each instance in the design. The excitation dependency graph shows how each entity could possibly trigger another entity directly.

There exists edges between various entities according to the following rules.

- Each word in the right hand side of a continuous assignment has an edge from it to all words in the left hand side of the same assignment.
- Each word in a port assignment has an edge to/from the respective ports of the child instance depending on whether the port assignment is input/output.
- An always block has an edge from itself to all the words that are defined with in the always block.
- A word in the sensitivity list has an edge from itself to the always block it is triggering.

The excitation dependency graph for the example Verilog code mentioned above is shown in figure III-A. The nodes are labeled as <module-name>_entity. *A possible recursive triggering of an always block is identified by a cycle in the excitation graph.*

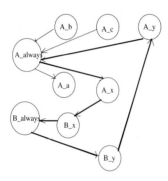

Fig. 2. Excitation Graph

B. Expression Generation for Recursively Triggering Always Blocks

We firstly introduce the idea of ternary model used in STE [2] and then present an extension of this model which we use to generate word level expressions. Randal.E.Bryant and C.J.Seger used the **Ternary Model**[2] to perform symbolic simulation over switch and gate level circuits. As an extension to the binary domain which contains the values $\{0, 1\}$, the ternary domain (**T**) contains the values $\{0, 1, X\}$. X is typically used to denote an unknown or an indeterminate value. A partial ordering relation is defined on the three values 0, 1 and X on the basis of "weakness in information content" denoted by $\sqsubseteq$. $\mathbf{0} \sqsubseteq X$ and $\mathbf{1} \sqsubseteq X$

Also, to make this partial ordering a lattice, we introduce a new value $\bot$ and define $\bot \sqsubseteq \mathbf{0}$ and $\bot \sqsubseteq \mathbf{1}$

A simple extension to the ternary domain (**T**) to allow for operation on words is shown in figure 3 . Let a and b be two words each of length n. The ith bit of a is addressed as $a[i]$. We define $a \preceq b$ if and only if :
for all i in 1 to n, $a[i] \sqsubseteq b[i]$ and
Any value $\sqsubseteq \bot$
Figure 3 summarizes the partial ordering relation ($\preceq$) on ternary and extended ternary models

Fig. 3. Ternary and Extended Ternary Models

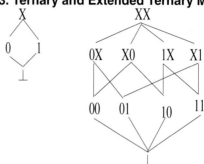

Indeed, an extended ternary domain for n-bits may actually be obtained doing a n-time product ($\mathbf{T}^n$) of the simple ternary domain. However, we prefer to use a smaller lattice by representing a set of values in $\mathbf{T}^n$ by a single value. In this case, the least upper bound of every element in the power set of $\mathbf{T}^n$ is used as a representative value for that element.

We assume that *each always block executes for a finite number of times between every two clock ticks.* This means that after a certain number of executions, the variables in the sensitivity list reach their fixed-points, ie., further executions of always blocks do not produce any change in the values assumed by these variables. (The formulation presented below however holds even if this assumption is violated.) We call an *epoch* as the uniform time period during which all signals stabilize in their values. For convenience, we may assume that each epoch aligns with a clock tick. We assume that each word takes values from the extended ternary lattice. The summarization of two values v_1 and v_2 in our domain is obtained by finding the least upper bound of both the values, **lub**(v_1, v_2).

Let us assume that there exists a module M with a single always block. Let $x_1, x_2, ..., x_n$ be the n variables in the module M. Let us symbolically execute the always block *once* with the initial values of the variables $x_1, x_2, ..., x_n$ to be $v_1, v_2, ..., v_n$, represented in vector form as $\overline{v}$. The new values for each variable after executing the always block once can be given as function of $\overline{v}$. Let $E_i(\overline{v})$ be the expression generated for the variable x_i by executing the always block once.

The new values for variables $x_1, x_2, ..., x_n$ are therefore $E_1(\overline{v}), E_2(\overline{v}), ..., E_n(\overline{v})$ respectively. This is represented in vector form as $\overline{E}(\overline{v})$. Let $\overline{I}$ be the vector of initial values for the variables in the current epoch.

Ideally, we would like to find the value of $\overline{E}^n(\overline{I})$ as n tends to infinity. However, since we donot know when it is going to converge, we keep summarizing the values a variable assumes during a simulation run. If the always block is executed zero or one time, its effect on a variable x_i is, given by, **lub**$(v_i, E_i(\overline{v}))$. In a similar way, we can capture the effect of executing the always blocks for zero or more number of times by repetitively applying the **lub** operator. We repeat this process until the application of **lub** reaches a fixed point. The expressions for the set of variables defined in a recursively triggered always blocks are therefore of the form :
lfp(**lub**($\overline{I}, \overline{E}(\overline{v})$))
where **lfp** is the least fixed point operator.

For the above formalism to be applied, we need that the expressions E_i satisfy the following :
1. $E_i(\bot) = \bot$
2. $E_i(\textbf{lub}(a,b)) \preceq \textbf{lub}(E_i(a), E_i(b))$

Both these properties are satisfied by the operators in Verilog since they are described only on the concrete binary domain.

A key point to note here is that the fixed point of the function above is obtained in a finite sequence of steps. This is because, each time we apply a **lub** operation over two operands, we either return one among the operands or move higher than the two operands in the extended ternary lattice. In the former case, we reached a fixed point. In the later case, we apply **lub** once again on the resulting value and some other value. As this process continues we are sure to hit a fixed point, because the lattice is a finite structure and once the topmost point is reached, applying **lub** would return the same value. Thus, in the extreme case, once could go up to the value of XXX and stop there. Termination of fixed point computation process in a finite time is indeed a desirable feature of using the extended ternary lattice.

The caveat however is that, by moving higher in the lattice, we are over generalizing the values and loosing information. In the worst case, we could end up saying that a variable assumes the value XXX which gives absolutely no information about the actual value of the variable. It should be noted that it is not always true that $\overline{E}(\overline{I})$ is $\preceq$ than $\overline{I}$. In other words, $\overline{E}(\overline{I})$ could be equally strong in information content as is $\overline{I}$. Hence applying a **lub** operation over these values can result in summarizing values too much. A simple enhancement could be that, we find $\overline{E}^k(\overline{I})$ for some value k and if a fixed point is not reached by then, we start summarizing using the **lub** operator.

From the above formulation, we get the stabilized values at the end of one epoch. For the variables that do not represent external inputs, the values at the end of this epoch serve as the initial values for the successive epoch. It should be noted that the function $\overline{E}$ is obtained by symbolically executing each always block in every module *once*. Thus, parsing through the CDFG of the design once should be sufficient to generate the required expressions.

We present below the expressions obtained by one time execution of always block for the example design described above. These expressions are actually the definitions for the function E_i presented in the above formalism. For a variable v in module M, let the initial value of the variable be M_v and the final value obtained after executing the always block once be M_v'. All the operators used in the expressions have the same meanings as those in Verilog HDL.
For module A :
A_x$'$ = (A_b==000) ? (A_c << 1) : A_x
A_a$'$ = (A_b!=000) ? (A_y & A_b) : A_a
For module B :
B_y$'$ = (B_x==100) ? (B_x << 1) : B_y
From the port connection statements we have that :
B_x$'$ = A_x$'$ and
A_y$'$ = B_y$'$
Therefore, by making the port assignments in order (input followed by output), we get the following values for the variables in terms of primary inputs :
For module B
B_y$'$ = (((A_b==000) ? (A_c << 1) : A_x)==100) ? (((A_b==000) ? (A_c << 1) : A_x) << 1) : B_y
For module A :
A_x$'$ = (A_b==000) ? (A_c << 1) : A_x
A_a$'$ = (A_b!=000) ? (((((A_b==000) ? (A_c << 1) : A_x)==100) ? (((A_b==000) ? (A_c << 1) : A_x) << 1) : B_y) & A_b) : A_a

C. Other Extensions

Edge Triggered Always Statements : To deal with edge triggered always blocks we propose a simple extension to the above approach. We assume that our domain has two more values ↑ (for a positive edge) and ↓ (for a negative edge). Also we introduce three new functions :
posedge(x) - returns true, false or indeterminate states
negedge(x) - returns true, false or indeterminate states

ite'(c,t,e,i) - evaluates condition c. If it is true, t is evaluated and returned. If it is false e is evaluated and returned. If it is indeterminate i is evaluated and returned.
As an example, consider the design:
always@(posedge clock)
begin
    out = in;
end
The expression for the new value of variable *out* may be given as : **ite'**(posedge($clock$),in,out,lub(in, out))

Ports and Continuous Assignments : According to Verilog semantics, continuous and port assignments can be made as soon as there is a change in the value of its operands. Since these assignments always represent equations that hold true in a stable state, we model each continuous assignment and port assignment as a constraint.

D. Comparison with Gate-Level Symbolic Simulation

The technique of symbolic simulation used in Symbolic Trajectory Evaluation (STE)[2] operates on the gate level model of a circuit. Besides the level of abstraction on which the simulation is carried on, there is also a difference in the semantics of the expressions generated by STE as compared to our approach. In STE, expressions are generated by *propagating* the input symbolic values into the circuit according to the circuit excitation functions. At any given instant of time, a snapshot of this simulation represents a set of possible scalar value simulations which can be obtained by instantiating the input symbols to various values.

On the other hand, our approach to simulation attempts to capture symbolically the local relations within a behavioral construct of the design. An actual simulation run on hardware corresponds to a particular ordering these local relations. Unlike in STE, we cannot get a valid simulation run by merely instantiating all the symbols to some values. It is necessary to impose an ordering among these expressions.

IV. Excitation Dependency Analysis

The fixed point formulation for expression generation leads to summarizing of values and hence may result in approximate outputs. We propose certain techniques to improve the accuracy of the approximations by separating those always statements which need a fixed point formulation from those that do not. Clearly, the always blocks that do not lie on a cycle in the excitation dependency graph do not need a fixed point formulation as they are executed atmost once in a single clock tick.

A naive approach to identify the recursively triggering always blocks is to first create an excitation dependency graph between various entities of the design as mentioned before. Once we obtain the initial excitation dependency graph, finding the transitive possibilities of excitation is simply obtained by finding the transitive closure of this graph. If after finding the transitive closure, we see that an always block excites itself, then it means that this always block can potentially re trigger and hence expressions are to be generated using the fixed point formulation. The caveat in this approach, however, is that this does not scale well with the size of the design. Assuming that there are n entities in the design, finding transitive closure in the worst case could be $O(n^3)$.

A. An Approximate Algorithm

We now propose an approximate algorithm which works much faster than the naive approach but with an approximation. This algorithm uses the distinction between the instance level view and module level view of the design. An *instance level view* would mean the picture that we get by instantiating every module according to each instance statement in the design. In this view, each instance(except the topmost one) has one and only one parent. On the other hand, in *module level view*, we instantiate each module only once and allow the module to have multiple parents if it is instantiated more than once. The following example clarifies the difference between both the views.

module A	module B(i,o)	module C(i,o)
.	.	.
B b(b_i,b_o)	C c1(c_i1,c_o1).	.
C c(c_i,c_o)	C c2(c_i2,c_o2)	**endmodule**
.	.	
endmodule	**endmodule**	

The instance and module views of the above design are shown in figure 4.

Fig. 4. Instance and Module views

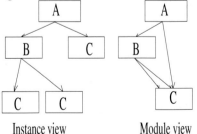

The approximate algorithm detects cycles by finding transitive closure on the module level view of the design.

Since each module is considered only once, we get a much smaller excitation dependency graph and hence transitive closure computation will be faster. From the properties of instance and module level views it can be proved that *All cycles in the instance level view indeed occur as cycles in the module level view*. However, some of the dependencies here are inexact. For instance, in the example presented earlier one could wrongly infer that variable c_i1 in module B can have an edge with the variable c_o in module A thorough module C which has multiple parents(both A and B). Thus, because of these stray dependencies one could arrive at flagging more always blocks to be recursively triggered than those that actually do. Also, after the always blocks that recursively trigger themselves in the module level view are identified, we assume that each instantiation of this always block in the instance level view is also recursively triggered. This is yet another approximation, because not all instantiations of this always block need to be on cycles.

Below is an example of how the module level view algorithm can infer cycles wrongly. Consider the following Verilog code.

module S	module $A(Ia,Oa)$
reg [0:9]Ia,Ib,Oa,Ob;	reg [0:9]Ia,Oa,Oca;
assign Ib = Oa;	C c_a(Ia,Oca);
A a(Ia,Oa);	**always**@(Ia or Oca)
B b(Ib,Ob);	Oa = Ia + Oca;
endmodule	**endmodule**
module $B(Ib, Ob)$	**module** $C(Ic, Oc)$
reg [0:9]Ib,Ob,Ocb;	reg [0:9]Ic,Oc;
C c_b(Ib,Ocb);	**assign** Oc = Ic << 1;
always@(Ib or Ocb)	**endmodule**
Ob = Ib + Ocb;	
endmodule	

Fig. 5. Instance Level View - No cycles

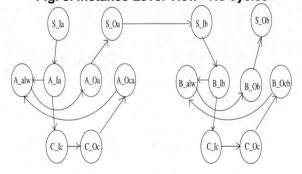

The excitation dependency graphs obtained by instance and module level views of the above example design is shown in figure 5 and 6. It can be observed that in the instance level view, one does not infer a cycle, but in the

Fig. 6. Module Level View - Inferring cycles.

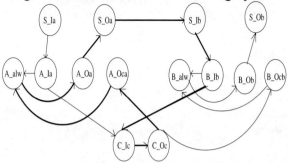

module level view, however, a cycle is inferred in the path B_Ib → C_Ic → C_Oc → A_Oca → A_alw → A_Oa → S_Oa → S_Ib → B_Ib.

Assuming there are k entities in the design at module level, the worst case time complexity of this algorithm is $O(k^3)$, since this is also a problem of finding the transitive closure.

V. Experimental Results

The simulation was run on parts of an example design called "**x86**". The **x86** project was an effort by the VLSI team at IIT Madras to create a public domain RTL synthesizable for the x86 CPU. Details of this project can be found at http://vlsi.cs.iitm.ernet.in/x86_proj/x86HomePage.html. Table II is a brief description of the complexity of the project at both instance level and module level views.

Table III shows the sizes of the largest expressions obtained by one time execution of always blocks of certain modules. These are the sizes of E_is as referred in section 3.2. Since these expressions are represented as DAGs, each common sub expression is counted only once. The number of common subexpressions as well as the tree sizes(expanded dags) for these modules is also presented.

Table IV compares the number of always blocks that actually exist in the design with the number that were detected to be on cycles by the approximate and naive algorithms. The naive algorithm on the complete x86 couldnot terminate before the timeout period of 6 hours when run on an IBM PC (Pentium 4 - 2.4GHz, 1GB RAM). In the case of simulating the complete x86, the size of the graph is 150,472 (which is the sum of the number of always blocks and the number of atoms in the instance view as shown in table II). Comparatively, the size of the corresponding graph used by the approximate algorithm

TABLE II. Module and instance views of x86

	Module View	Instance View
No of instances	122	12,111
No of always blks	123	357
No of words	5,888	150,115

TABLE III. Word level expression sizes

Module	Word	Size	No of common sub-exps	Tree size
conUnit	infp2[0:79]	938	34	29305
conUnit	buffer[0]	762	6	1646
conUnit	dataWr[0:31]	1449	86	4102
conUnit	address[27:35]	1184	61	24598
myAlu	cla1op1[0:31]	1472	4	2700
myAlu	twoscompIn1[0:31]	972	3	1739
myAlu	zfin[0:31]	1890	32	4608
myAlu	Result[1:3]	3524	4	6390
Decoder	MemOpEn[0]	964	6	1309
Decoder	ARPLEn[0]	678	2	921

TABLE IV. Detecting always blocks on cycles.

Module	#always blks	#detected on cycle	
		Approx Algo	Naive Algo
full x86	357	221	-
multiply32	33	33	33
Decoder	11	11	11

has only 6011 ndoes. It is worthnoting that, despite a huge reduction in graph size by the approximate algorithm, it was still able to infer that 136 *always* blocks are not on cycle. In the decoder unit each module is instantiated only once and hence, the approximate algorithm performs the same as the naive algorithm.

VI. Conclusion

Our focus of current research is on building up of a constraint solver which can reason on lattice operators introduced by the fixed point formulation. SMT solvers which operate on boolean as well as bit vector domains are being studied. The expressions that the simulator generates are yet large and it is necessary to optimize or approximate them for the idea of word-level symbolic simulation to be made applicable on large designs such as microprocessors.

Acknowledgment

The authors would like to thank Ashutosh Kulkarni and Kaustubh Nimkar for their inputs and help in the implementation of the symbolic simulator.

References

[1] R. E. Bryant, "Symbolic simulation - techniques and applications," in *Design Automation Conference*, 1990, pp. 517–521. [Online]. Available: citeseer.ist.psu.edu/bryant90symbolic.html

[2] R. E. Bryant and C.-J. H. Seger, "Formal verification of digital circuits using symbolic ternary system models," in *Proc. 2nd International Conference on Computer-Aided Verification (CAV'90)*, E. M. Clarke and R. P. Kurshan, Eds. Springer-Verlag, 1991, vol. 531, pp. 33–43. [Online]. Available: citeseer.ist.psu.edu/643147.html

[3] C.-J. H. Seger and R. E. Bryant, "Formal verification by symbolic evaluation of partially-ordered trajectories," *Formal Methods in System Design: An International Journal*, vol. 6, no. 2, pp. 147–189, March 1995. [Online]. Available: citeseer.ist.psu.edu/seger93formal.html

[4] M. Pandey, R. Raimi, D. L. Beatty, and R. E. Bryant, "Formal verification of powerpc arrays using symbolic trajectory evaluation," in *Design Automation Conference*, 1996, pp. 649–654. [Online]. Available: citeseer.ist.psu.edu/article/pandey97formal.html

[5] Z. Z. et al., "Functional test generation based on word level sat," in *Journal of Systems Architecture*, 2005, pp. 488–511. [Online]. Available: citeseer.ist.psu.edu/su96automatic.html

[6] C. Wilson, D. L. Dill, and R. E. Bryant, "Symbolic simulation with approximate values," in *Formal Methods in Computer-Aided Design*, 2000, pp. 470–485. [Online]. Available: citeseer.ist.psu.edu/wilson00symbolic.html

[7] J. X. Su, D. L. Dill, and C. W. Barett, "Automatic generation of invariants in processor verification," in *First international conference on formal methods in computer-aided design*, M. Srivas and A. Camilleri, Eds., vol. 1166. Palo Alto, CA, USA: Springer Verlag, 1996, pp. 377–388. [Online]. Available: citeseer.ist.psu.edu/su96automatic.html

[8] P. Cousot and R. Cousot., "Abstract interpretation: a unified lattice model for static analysis of programs by construction or approximation of fixpoints," *Sixth Annual ACM SIGPLAN-SIGACT Symposium on Principles of Programming Languages*, pp. 238–252, 1977. [Online]. Available: citeseer.ist.psu.edu/bryant86graphbased.html

[9] I. std 1364-2001, *Verilog Hardware Description Language*, IEEE Computer Society.

System Level Validation and Test
MTV 2007

Top Level SOC Interconnectivity Verification using Formal Techniques

Subir K. Roy
Texas Instruments India, Bangalore
(subir@ti.com)

Abstract

SOCs are designed by integrating existing in house cores/intellectual properties (IPs), or third party core/IPs provided by external vendors to reduce design turn-around time and cost. The integration process in realizing an SOC implementation consists of several different kinds of integration which can be classified as (1) static or non-functional integration, consisting of simple electrical connections (or hookup) of the inputs and outputs of different IPs, and (2) dynamic, or functional integration, where, besides the pure electrical connectivity, a temporal and a functional dimension needs to be taken into account. Given the size of present generation SOCs there is ample scope of inadvertent design errors being introduced during the integration process. It has been observed in many in-house SOC designs that a large percentage of these errors (80%) are contributed by pure connectivity errors. Severity of implications of some of these errors, are dependent on when they are detected in the design verification cycle, and on how easy it is to correct them in the implementation. In this paper, we present the challenges involved in detecting such integration errors in complex SOCs through the use of formal verification techniques. The main contributions of this work are (1) effective use of formal techniques based on symbolic model checking in the top level verification of SOC integration, (2) effective use of abstraction and modeling of SOC sub-systems in enabling assertion based formal verification, (3) automated generation of assertions and constraints to detect integration errors, (4) automated generation of scripts to capture the SOC design information and invoke a formal verification tool on which to prove the validity or correctness of these assertions, and (5) case studies from different categories of SOC integration to highlight the benefits of the proposed approach. These techniques have been applied to verify different kinds of integration described above in several SOCs designed in Texas Instruments, with promising results.

1. Introduction

Present generation SOCs are designed with a strong focus on re-use of existing in house cores/IPs, or third party IPs provided by external vendors to reduce the design turn-around time and cost. A core/IP is a pre-designed and a pre-verified Silicon circuit block. Different IPs typically used in SOCs consists of, CPU cores, DSP cores, memory cores, power management cores, analog and mixed signal cores (DAC, ADC and PLL), different peripheral interfaces, DFT and BIST IPs, I/O pads and custom logic blocks specific to a design. These cores/IPs conform to different functional (MPEG, JPEG, CRC, etc.), interconnect/bus (PCI, SCSI, USB, 1394, IrDA, Bus Bridges, etc.) and networking (10/100 ethernet, ATM, etc.) standards, as may be applicable to them.

Given the size of present generation SOCs and the different approaches taken to implement them through a team of designers, there is ample scope of inadvertent introduction of design errors. Some of these errors arise from one or more of the following reasons, such as, incorrect specifications, misinterpretation of specifications, misunderstanding between designers, missed cases (many a times the common communication interface between the system designers and the hardware designers is based on an imprecise natural language document), protocol non-conformance, resource conflicts, and cycle-level timing errors. These introduce different classes of design errors [1], such as, errors in interactions between IPs/cores, conflicts in accessing shared resources due to incorrect arbitrations, dead-locks, live-locks, priority conflicts in exception handling and un-expected hardware software sequencing. Introduction of these design errors and the inherent incompleteness of the different design views seen by an IP developer and an IP integrator in the SOC context renders the top level verification of SOCs a difficult problem. While the IP developer has a detailed understanding of the IP micro-architecture, he/she has an incomplete view of the target SOC. In contrast, while the SOC architect (or the IP integrator) has a very good understanding of the intended architecture, she/he views the IP as a black box, thereby, exercising very little observability, or controllability at the IP ports during the verification process.

Severity of implications of some of the above issues, are dependent on when the errors are detected in the design and verification cycle, and how easy it is to correct these errors in the implementation. For example, specification errors, even if they are simple, are susceptible to late detection in the entire design and validation flow, and manifest themselves in erroneous corner case temporal behavior, which are difficult to pin-point, and correlate to the contributing IP. Some of the implications of this on the verification of an SOC are that, each individual SOC IP/core needs to be separately verified rigorously and exhaustively, and that the entire SOC itself needs to be verified extensively. To achieve this requires, efficient verification methodologies based on re-use of verification components of individual cores/IPs through automation, abstraction of verification goals (for example, signals to transactions, end to end transactions), efficient verification tools and a very high level of automation, wherever, possible in the entire verification process.

Top level SOC integration can be classified, as follows, 1). Static or non-functional integration consisting of simple electrical connection (or hookup) of the inputs and outputs of different IPs present in the SOC, and 2). Dynamic, or functional integration, where besides the pure electrical connectivity a temporal and a functional dimension needs to be brought into the picture. An example of the temporal dimension is that of pipeline registers being added into data-flow paths to meet timing constraints, while an example of the functional dimension is the switching between the functional mode and the test mode of operation.

Towards achieving the above objectives, we have taken a complementary approach of deploying both, dynamic and static verification methodologies based on simulation and formal model checking, respectively. In this paper, we report our efforts in deploying formal verification to the task of top level SOC verification through case studies, highlighting various aspects and phases of top level SOC verification. Our justification of using formal approaches wherever possible can be found in [2]. More specifically, we present the challenges involved in detecting such integration errors in complex SOCs through the use of formal verification techniques. The main contributions of this work are (1) effective use of formal techniques based on symbolic model checking in the top level verification of SOC integration, (2) effective use of abstraction and modeling of SOC sub-systems in enabling assertion based formal verification, (3) automated generation of assertions and constraints to detect integration errors, (4) automated generation of scripts to capture the SOC design information and invoke a formal verification tool on which to prove the correctness of these assertions, and (5) case studies from different categories of SOC integration to highlight the benefits of the proposed approach. These techniques have been successfully applied to verify different kinds of SOC integration, in several SOCs designed in Texas Instruments, with promising results. Experimental data based on a commercial formal verification tool [4] show that the proposed approach is an order of magnitude faster than approaches based on simulation.

The organization of the paper will be as follows. In Section 2, we briefly review symbolic model checking based formal verification. In Section 3, we present our methodology for the first classification and also present a case study on which this was successfully applied. The case study describes the top level SOC interconnectivity verification using formal techniques carried out on a very large SOC designed in-house (presented here in the Appendix section). In Section 4, we outline our methodology for the second classification as applied to the problem of verifying the correctness of I/O pad frame and functional pin multiplexing. We present data from an in-house SOC design where our automated methodology is being successfully applied and has replaced the existing simulation based approach completely. In Sections 5, 6, 7, 8 and 9 we present interesting case studies, again, pertaining to the second classification drawn from the domain of SOC DFT logic – verification of its insertion and integration into SOCs using formal techniques. Some of our efforts on SOC integration verification of BIST DFT logic blocks can be found in [2]. More specifically we will present the formal verification of several non-BIST DFT logic blocks, such as, the test mode controller, test sequence setup for different test modes, test pin multiplexing and boundary scan, and in circuit emulation and test IP, respectively [3]. These together constitute a major portion of DFT logic integrated into any SOC designed in Texas Instruments. In Section 10, we present our views on the issues and challenges that were encountered in adopting the formal approach to the problem of verification of SOC integration. We present our conclusions in Section 11.

2. Review of Formal Verification based on Symbolic Model Checking

In model checking, a mathematical representation of the design in the form of a finite state machine (FSM) is first constructed. A desired behavior is then stated in terms of a property (or a specification) in temporal logic. The FSM is then analyzed using state traversal techniques, starting from the set of initial states, to check whether it satisfies the temporal property on all, or at least, one path of the state transition graph that is implicitly generated by the above state traversal. This state traversal is known as *reachability analysis*. In case the temporal property is violated, a trace with respect to the primary inputs and state variables of the FSM, starting from its set of initial states, is generated up to the Kth set of states, where the property fails on one of its states. This is known as an *error trace*. It is possible to generate this error trace because, the collection of states that are reachable on every clock cycle, starting from the set of initial states, is stored internally by the model checker. When each of the reachable state set is implicitly represented as a *binary decision diagram (BDD)*, the model checking technique is known as symbolic model checking (SMC). In SMC, properties are specified using different temporal logic, e.g. *Linear Temporal Logic (LTL)*, or *Computation Tree Logic (CTL)*, or *Property Specification Language (PSL)* [8,9].

The formal verification of a hardware module begins by stating a formal property and then checking that the hardware module satisfies this property. Invariants are the most commonly specified properties. Invariant properties express conditions on a hardware module that should never happen in a reachable state. Conversely, invariants can also be conditions that should always be true in a reachable state. An invariant property is a Boolean formula over the signals of a module. A module M satisfies an invariant property I if every reachable state of M satisfies I. Thus, invariant verification is performed on a module by computing its set of reachable states. However, this computation is difficult because the set of reachable states can be exponential in the number of input and internal signals in the module. This exponential growth in the number of states is known as the *state explosion problem*.

The state explosion problem in SMC limits the size of designs that can be formally verified. A widely accepted thumb rule is that modules should have less than 500 flip-flops for SMC to succeed. This necessitates carrying out partitioning and abstraction of the design for the formal approach [9].

3. Case study of a top level SOC interconnectivity verification using formal verification:

As an example of the first classification related to top level SOC interconnectivity verification, i.e., static or non-functional integration consisting of simple electrical connection (or hookup) of the inputs and outputs of different IPs present in a SOC, we present a case study of an in-house SOC designed in the recent past.

Based on several SOCs designed in-house, it was found that it is the initial phase in the design cycle, involving integration of different IPs/cores, which contributes a significant amount to the design time, with most of this time being devoted to weeding out the errors in design related to the first category of integration. To pinpoint these errors requires time consuming complex simulations runs based on implicit functional top level test benches, where simple connectivity errors between IPs are detected, only after these manifest themselves as erroneous top level SOC output patterns. It is obvious that this will require enormous effort, as the designer will need to correlate erroneous top level behaviour with IP level boundary behaviour. This involves enhancements and modifications to existing top level test benches, to include IP level boundary behaviour, and additional test cases to enable detection of each connectivity error. The other important contributing factor is the fact that the top level test benches are applicable only after each IP in the SOC has been fully integrated in the implementation RTL. This implies that the integration verification begins fairly late in the design cycle, resulting in additional effort needed to resolve integration errors, as multiple IP level connectivity errors may manifest themselves in a single top level SOC error.

These errors arise due to two reasons related to the automated generation of the SOC implementation RTL from the connectivity specification provided as a spreadsheet table. The first reason being attributable to the software wrongly implementing the automated generation of RTL, and the second related to connectivity specification errors introduced by designers. Besides the connectivity errors, specification errors and logical errors are the other two contributing factors. Typical distributions observed for these errors are as follows, 84% contributed by connectivity errors, 10% contributed by specification errors and 6% contributed by logical errors, respectively.

Given the above facts, a formal verification approach was taken to address the issue of 1) detecting all connectivity errors in every level early in the design cycle (as soon as, two or more IPs were connected), 2) to reduce the over all SOC interconnection verification time and effort by focusing on the core issue through automated connectivity property generation for formal verification, and exploiting the powerful error trace generation capability available in formal verification to quickly pin-point and correct errors.

The proposed flow is shown below in Figure 1. In this flow we separate the connectivity specification needed by the software to generate the RTL, from that needed by the script to automatically generate the connectivity properties needed by formal verification. Two different designers capture these specifications in Excel spreadsheets. We use PSL to specify the properties and IFV (a symbolic model checking tool from Cadence [4]) to validate these properties formally.

The SOC used in the case study has a total of 42 different IPs and 117 instances of these IPs at the top level. The results presented below are for a sub-system in the SOC.

1. Total connectivity PSL properties for sub-system = 6500
2. Time taken to validate PSL properties in IFV = 5 hours (3 to 4 seconds per PSL property) on a 64 bit Linux platform with 4GB RAM.
3. Projected gain vis-à-vis simulation is 35X, when applied to the whole SOC.

The key advantage of the proposed approach is that, it can be applied to a partially complete RTL with minimal effort, as the connectivity properties and the script needed to run them on IFV are auto-generated.

4. Formal Verification of IO Pad-Frame Logic and Functional Pin Muxing

As another example of the first classification, we present the case study of verification of IO pad-frame logic and functional pin muxing using formal techniques. IO pad-frame logic is a layer of logic present between the core logic and the input, output and bi-directional pads in a SOC. This is shown in Figure 2 below. It essentially allows sharing the costly chip package input output pin resources with multiple internal core logic data and control signals. The verification of the simple muxing logic constituting the IO pad frame logic can consist of proving the correctness of several thousand different connections of core logic signals to the package input, output and bidirectional pins under different modes of operation in both the functional modes and the test modes of operations. Using top level simulation test benches it may not be possible to exercise the degree of controllability and completeness needed to ensure total absence of bugs, for the same reason as highlighted in the earlier section. We developed an automated FV based verification flow which completely removed the effort of manual generation of the formal properties. This flow has completed replaced all simulation based flows that were used earlier and has been validated on a large number of in-house SOCs.

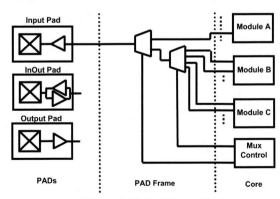

Figure 2. IO Pad Frame Logic

Table 1 below gives the results of an ongoing in-house SOC design undergoing revisions in the design of the IO pad frame logic.

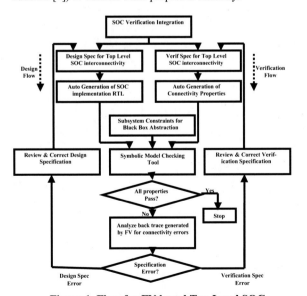

Figure 1. Flow for FV based Top Level SOC Interconnectivity Verification

5. SOC Memory Data Path Verification (MDP) Using Formal Techniques

As an example of the second classification related to top level SOC interconnectivity verification, i.e., dynamic, or functional integration, where besides the pure electrical connectivity a temporal dimension needs to be brought into the picture, we consider the integration of programmable memory BIST controllers into a SOC.

Property Class	Total Properties	# of Passing Properties	# of Failing Properties
Input	130	130	0
InputProhibit	66	66	0
Output	160	75	85
PadOutLogic	109	67	42
PadOutEnLogic	109	98	11
PadPIPU	100	98	2
PadPIPD	100	98	2
PadHoldL	238	236	2
PadHoldH	238	208	30
PadHoldZ	238	206	32
PadHHVZ	113	113	0

Table 1. FV run statistics of IO Pad Frame properties

Hardwired controllers target test pattern generation algorithms for detecting known memory faults, while in a micro-coded controller the test algorithm is programmable. It is possible to load and generate pattern sequences, post silicon fabrication, to detect hitherto un-modeled and unknown memory faults arising in a new fabrication process. These are therefore, referred to as programmable memory BIST, or PMBIST/PBIST controllers [5].

As design and integration technologies mature, newer defects and fault models are introduced. This is very well illustrated in the field of memory testing in terms of the growing amount of memory being integrated into large SOCs, the complexity of testing them, and the increasing gamut of solutions available totest them [6,7]. Details of formal verification of hardwired memory BIST, PMBIST/PBIST controllers through the formal verification of many generic blocks present in them and their integration into SOCs, can be carried out using techniques given in [2]. We briefly discuss below the integration verification using formal approaches.

Figure 3 above shows the block diagram of a PMBIST controller, while Figure 4 shows its integration in a SOC, where it is selectively connected to an embedded memory in a given sub-system in the SOC through the global memory data path (MDP) and the sub-system memory data path blocks. Figure 5 shows the different pipeline depths that different embedded memories encounter in different sub-systems within the SOC. Verification of the integration of the PMBIST controller essentially involves verification of the above MDPs. Formal properties can be broadly classified as three types : 1) for broadcast signals - signals like address and write-data launched at the PMBIST boundary should reach the memory boundary after 'forward path latency' cycles; 2) for control signals – a) the control signal is asserted at the memory boundary after 'forward path latency' cycles only if all the gating conditions for this signal assumes correct values, b) if any one of the gating conditions for this signal is not true or does not assume a correct value, then, it should remain de-asserted at the memory boundary; 3) for return data from memory – return data from the memory should reach PBIST boundary after 'return path latency' cycles, if two control signals are properly selected. The formal properties are auto generated on a per memory basis using memory information captured in a spread sheet. Depending on the type and ports of each memory, properties will be generated to verify the connectivity from PMBIST boundary to the memory boundary.

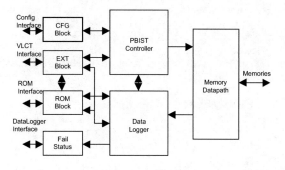

Figure 3. PMBIST Logic

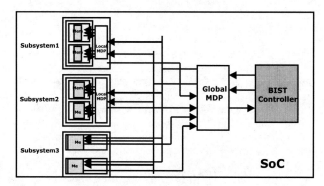

Figure 4. PMBIST Controller Integration in a SOC

In many TII SOCs, a deeper look into the functionality of MDP reveals that most of the modules are redundant

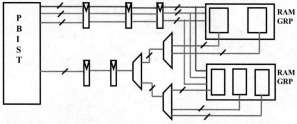

Figure 5. Pipeline registers in a PMBIST memory data path

in the context of interconnection of the PMBIST controller to the memories and associated logic. Thus, interconnect and associated logic needs be formally verified by suitably black-boxing such redundant modules. It is also imperative to ensure that the same clock from the PMBIST controller that drives the embedded memory also drives the pipeline registers that lie in the forward and the return paths between the PMBIST controller and the memory. To address this issue through FV it becomes essential to include the clock sub-module from the PMBIST controller. It is also essential to model the one flop delay introduced by the PMBIST controller on a selected clock before it is sent out as its primary output, and to model the data read latency introduced by a memory. A simple solution to this issue, which was reported as a bug in one of the in-house TII SOC design, is to carry out the above tasks in the verification modeling layer to enable script based auto-generation of the relevant properties. By knowing all the clock sources from the top level module, properties for the interconnectivity and associated logic can be written to verify their correctness with respect to their specified clock coming out from the clock multiplexing logic of the PMBIST controller. Information related to the clock source and the selection line input to the clock multiplexing logic in the PMBIST controller are captured by the SOC designers in a spread sheet. We developed an efficient methodology to formally verify such interconnects and their associated logic. The process of property generation has been automated given the hierarchical paths in the RTL of the memories and the PMBIST controller along with the top level constraints that are required to enter the appropriate test mode. This methodology has been applied to several in-house SOC designs. Results obtained are extremely promising from the point of view of catching bugs, as well as, completing the full verification with bug detection and elimination, requiring a turn around time which is much lesser than that possible through simulation. This approach revealed a fairly large number of critical bugs in different RTL releases in all the SOCs as highlighted in Table 2.

	No. of Mem	Property - Total		Property - Pass		Property - Fail		CPU Time (sec)	
SIMCOP	25	175	175	160	150	15	25	999.06	1261.88
TPDMA	6	30	30	30	24	0	6	222.45	222.45
VIMCOP	68	406	406	269	337	137	69	26822.50	28116.50
ARM	27	189	196	33	156	156	40	2891.87	1524.68
VPSS	20	287	301	275	250	12	51	2535.25	2083.27
XHPI	1	7	7	7	6	0	1	22.42	25.42
IMAGEROM	1	5	5	5	4	0	1	19.64	22.92

Table 2. Comparative Statistics of PMBIST MDP properties after FV run on different RTL releases of a SOC.

6. Formal Verification of In Circuit Emulation and Test IP (ICEPick)

SOCs typically have multiple sub-systems, each of which has a JTAG test access port (TAP) embedded in the core. The IP module manages these TAPs. At its root, it is a scan-path linker that allows the scan controller to selectively choose, dynamically and at runtime, the subsystem TAPs that need to be accessed through the device level debug port. The IP is configured via its IEEE 1149.1 JTAG TAP Controller. It supports serially linking up a parametrizable number of TAP controllers, individually selecting one or more of the TAPs for scan without disrupting the state of the other TAPs.

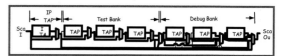

Figure 6. TAP linking in In-Circuit-Emulation and Test IP

Power management techniques employed within a SOC may cause power to be turned off in one of its subsystem, resulting in its unpowered TAP being part of a serially linked scan. The IP overcomes this by dynamically configuring the SOCs scan path so that it can be routed around the powered down subsystem to maintain scan path integrity. The IP in conjunction with other system and debug components is designed to support the following debug activities, 1) synchronized execution control of multiple cores, 2) debug while one or more modules are not powered, and 3) debug while one or more modules have a slow, or no clock. In addition, it supports a number of test activities, 1) chip level JTAG boundary scan, without the need for additional hardware jumpers, 2) running of production test vectors on a particular module, and 3) providing access to test TAPs within modules, or at the chip level.

To formally verify the IP, we sub-divided the task into verification of the following functionalities, 1). TAP Linking, 2). Registers, 3). JTAG Signals, 4). Scan Mechanism, 5). Secondary TAP Control, 6). Emulation Triggers, 7). Reset Module, and 8). TAP State Machine. Each of the functionality has been verified independently by taking as constraints the behavior of the rest of the modules. We prove these constraints while verifying respective modules. We briefly describe below the approach followed to formally verifying the TAP linking functionality.

Some of the properties to verify the TAP linking functionality are, 1) when the *Always-First* mode is used, ICEPick's TAP is always linked and visible between the device TDI and TDO, 2) when the *Key_Sequence* mode is activated the IP TAP is disconnected, 3). When the *Disappear_Forever* mode is activated the IP TAP is not visible until the next power on reset. Several JTAG TAP interfaces are controlled by the IP, namely, device level TAP interface, IP TAP interface and several secondary TAP interfaces as shown in Figure 6. The following functionalities in the context of TAP Linking are verified, TAP Ordering, TAP Numbering and TAP Connectivity. These are exhaustively verified by automatically generating properties capturing all possible combinations of visibilities of the TAPs.

7. Formal Verification of Test Mode Controller (TMC) Logic

The test mode controller (TMC) logic is an IP in the DFT logic suite which enables selection of different test-modes in a SOC and generation of control signals for the corresponding test modules associated with a test mode. TMC can be programmed through standard interfaces, such as, IEEE1500 or JTAG. Instructions are decoded and control signals are sent to enable or disable different test modes. For example, TMC based on the IEEE1500 interface, enables test engineers to program and control desired functionality by writing instruction and data into specific registers under the control of a finite state machine, which are then decoded to generate control signals. A brief description of its functionality is given with reference to Figure 7.

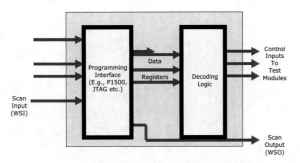

Figure 7. Test Mode Controller

Instructions are written through the scan input pin (WSI) and are shifted into the Instruction-Shift-Register (*shiftwir*) in the Instruction-Shift state of the FSM and are updated into the Instruction-Register (*updatewir*) in the Instruction-Update state. Data is shifted into the Data-Shift-Register (*shiftwdr*) in the Data-Shift state. Depending upon the instruction present in the Instruction-Register the proper Data-Register (*updatewdr*) is updated in Data-Update state. Control signals generation after decoding can be through three different ways. These are, 1). Output control signals could be connected to a bit of a data register. Simple connectivity checks suffice for such control signals, 2). Control signals are the output of a multiplexing logic selected by part-select of a register, and 3). Control signals are outputs of complex AND-OR driving logic. TMC was verified formally by first verifying the IEEE1500 interface, separately, through properties written manually, and then verifying the decode logic by generating its properties through an automated flow.

8. Test mode verification using FV for scan chain based test mode entry sequences

In case of a dedicated processor being present in the SOC it is possible that the test modes for both the SOC and the processor needs to selected through different test mode controllers based on multiple IEEE1500 controllers. In such cases, the ICEPick module discussed above, is used to select between multiple IEEE1500 controllers. This involves programming the ICEPick connect sequences to get into any test mode. One way to access the test modes is that the SOC and the processor IEEE1500 controllers can be treated as JTAG TAP controllers and connected between TDI and TDO along with other TAP controllers by selecting them by programming the corresponding bits in the data register of ICEPick TAP select instruction. All the selected TAP controllers and P1500 controllers will be connected in a daisy chain fashion. Instruction registers of all the selected controllers will be concatenated and a large instruction will be seen at the SOC level that can be programmed in a single IR shift and update sequence. All the TAP controllers and P1500 controllers can remain active at the same time in this mode of operation. The other simpler way to access the test modes would be to select the P1500 controllers individually so that they can be accessed directly through the SOC pins by programming the ICEPick

appropriately. None of the TAP controllers excepting the ICEPick TAP will remain active in such a mode of operation. This enables the test mode access to be much simpler.

It is, thus, clear that to validate the correctness of the different test modes through the test mode controllers involves firstly checking that the specified test mode entry sequences corresponding to different test modes are properly executed so that the appropriate decoding registers are written correctly with the desired values corresponding to each test mode. After this it is a simpler verification task to prove that the different register values generate the correct control signals relevant to a particular test mode. We tried two approaches to validate the correctness of the test mode entry sequences. In the first approach the activation and de-activation of the different test modes and the functional mode of operation were based on the test mode sequence entries being directly programmed through the primary ports of the SOC with the entire hierarchy of the SOC being taken into consideration. Formal properties in PSL were written with respect to the primary input and output signals of the SOC, in addition to all the constraints required for verification. This approach, ran into the well-known state-explosion problem present in symbolic model checking, as a large number of registers were present in the fan-in cone of different signals involved in all the properties with most of these registers being contributed by both the ICEPick module and the SOC JTAG register modules; besides, the properties also involved huge sequential depths.

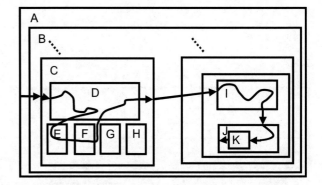

Figure 8. SOC Top Level View for Test Mode Entry Sequence

A deeper look into the functionality reveals that most of the portions of a SOC are redundant in the context of interconnection of the input ports of the SOC to the test mode registers and associated logic in the ICEPick module and SOC JTAG register modules. Thus, the verification task based on formal methods can be simplified by suitably black-boxing redundant portions of the SOC and using a compositional approach to contain the state explosion problem. This constituted our second approach, which we discuss in some details below. Apart from the correctness of the interconnectivity and its associated logic, it is also imperative to ensure that the same clock from the top level SOC port drives the ICEPick module which lies in the path between the SOC top level ports and the test mode JTAG registers.

Figure 8 describes the SOC hierarchy involved in the test mode entry sequence validation. Block A represents the topmost hierarchy of the SOC. Block B is a sub-system deep down in the mid hierarchy containing the ICEPick controller module which is represented by the block D. Blocks E, F, G and H represent the TAP, the clock voting, the trigger and reset modules, respectively. Blocks I, J and K representing the SOC JTAG register module, the DFT module and the test mode register modules respectively, are present in some other sub-system within the SOC. The trace of the path taken to program the test mode JTAG registers are as shown in the same figure.

Block Level Connectivity	Total Properties	Average Flip-Flops	CPU Time [Min]
ICEPick IP	61	170	38
JTAG Regs	10	90	20
Connectivity	14	2	3

Table 3. Statistics of Test Mode Entry Sequence properties

In the second approach to prove the above top level properties related to the validation checks associated with the activation and de-activation of various test modes, we took the approach of breaking them into the properties related to simple connectivity checks and properties related the correctness of sub-functions of the ICEPick and JTAG register modules. The latter properties were re-used from the suite of properties already developed for these individual modules as a part of our DFT IP verification efforts based on formal verification. The formal verification task for the second approach was broadly divided into the following categories, a) proper connectivity from top level SOC boundary to the ICEPick module boundary, i.e. ports of A to ports of C in Figure 8. b) functionality checks of the ICEPick module (block C), which was further broken down into the functionality checks of the ICEPick controller, the tap functionality and the clock voting logic, c) based on the values of ICEPick controller registers, connectivity of output signal at the ICEPick module boundary to the inputs of the JTAG register module boundary, d) functionality of JTAG register module (block I), which functions as a decoding logic for the various test modes, and e) proper connectivity from JTAG register module outputs to the test mode registers in the DFT module (port of I to port of J, and block K).

Using this approach none of the properties in any of the above categories encountered state explosion. Moreover, most of the properties were proven within reasonable time as reported in Table 3.

9. Formal Verification of Test Pin Multiplexing Logic

To reduce the number of pins in an SOC, some of the system pins are multi-plexed to function as either core pins, or as test pins. The TPM module reconfigures these functional pins to operate as test pins and carry test signals to the core DFT logic modules. Under a particular test mode, these test pins could be configured as input, output or bi-directional pins. Figure 9 gives the block diagram of TPM.

Boundary scan (BScan) cells associated with each system pin can be one of the standard boundary scan cells defined by the IEEE1149.1-2001 standard depending on the design requirements, with each of them serving a specific purpose and their choice depending on whether the system pin is input, output or bidirectional. The set of properties to validate the functionality of TPM and boundary scan cells can be automatically generated by a script. Some of the properties are, 1) connectivity between BScan cells and different input (output) pins of different test modules selected under different test modes, 2) write-enables of all the pins should be in high-impedance state if *CATSCAN* mode is selected, 3) If the *CATSCAN* mode is not selected, while the *SCAN_ENABLE* mode is selected, then the output enables of all pins other than scan-out pins should be in high impedance state, and the write enables of all the scan-out pins should be in the output mode, and 4) the output of the boundary scan cells should be connected to the corresponding functional pins if none of the test modes and the boundary scan mode is selected.

IEEE standard defines some specific functions which can be performed by boundary scan cells. These functions are described as instructions, such as, SAMPLE, PRELOAD, INTEST, EXTEST. Each standard boundary scan cell supports a subset of the available instructions. For example, BC_7 cell supports instructions SAMPLE, PRELOAD, EXTEST. Selection of a particular BScan cell depends on the desired functionality required in the corresponding system pin. For example, the SAMPLE instruction when invoked loads the system snapshot into the boundary scan cells which is then shifted out of the boundary scan chain. To validate this instruction the following functionality needs to be captured as formal properties, 1). On-chip Logic should be unaltered, 2).System snapshot taken in Boundary Scan cells in Capture Controller state, and 3).Captured snapshot shifted out in Shift Controller state.

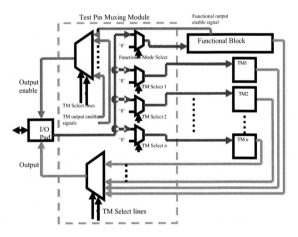

Figure 9. Block diagram of a typical TPM module.

10. Issue and Challenges

Our effort in deploying FV for DFT logic integration in an SOC has provided some useful insights into some of the issues and challenges. These are related to the inherent limitation of the formal verification approach based on symbolic model checking. While simulation can handle entire designs, state explosion in model checking limits designs that can be verified to less than 500 state bits. From a technical perspective, this potentially impacts verification of IP logic and its integration in many different ways which impacts its adoption by designers and the verification engineers. The main challenges that were faced involved finding automation routes to property generation, to abstraction of the modules and sub-systems not needed for a particular verification task, and to setting up appropriate environmental constraints for the un-abstracted portion involved in the verification task.

Another issue faced relates to short comings in the property suite being developed which can arise from many different factors, such as, 1) too few properties (has enough of the behavior been covered?), 2) inadequate or ambiguous specification ("A bug is actually a feature" and vice-versa), 3) inadequate sanity checks (while FV generates counter examples only for failed properties, none are generated for properties which pass, necessitating the need to ascertain that these are true passes), 4) over constraints and wrong assumptions, 5) over abstraction or simplification (incomplete understanding of implementation RTL), 6) miss out on verifying assumptions as guarantees, and 7) ignore multiple clock domains. This ties in closely with the issue of coverage and the adequacy of the set of properties being developed. Our approach has been to write exhaustive set of properties so as to subsume all of the specified behavior and verify it under the least possible environmental constraint.

11. Conclusion

This paper presented modeling techniques and approaches that can be taken to formally verify the integration of different IPs (both functional and DFT logic IPs) that are being increasingly used in SOCs to reduce design turnaround times and verification costs. It is shown how various integration tasks can be formally verified using suitable abstraction and partitioning. Improvements in verification quality and time highlight the comprehensiveness of formal techniques over simulation techniques and their potential to be the mainstream approach for verifying top level interconnectivity in SOCs.

12. References

1. T. L. Anderson, "Accelerating Bug Discovery with White-Box Verification", Proceedings of DAC, 2000.
2. Subir K. Roy and R. A. Parekhji, "Modeling Techniques for Formal Verification of BIST Controllers and Their Integration into SOC Designs", Proceedings of the VLSI Design Conference, Bangalore, 7th – 11th January 2007.
3. Subir K. Roy, Anindya S. Nandi and Bijitendra Mittra, " Embedded Tutorial : Formal Verification of DFT Logic and their integration in System on Chips – Practices, Issues and Challenges", 11th IEEE VLSI Design and Test Symposium, Kolkata, India, Aug 8 -11, 2007.
4. IFV – Incisive Formal Verification Tool, Cadence Design Systems Inc., 2005.
5. D.Gizopoulos, (Ed.), *Advances in Electronic Testing: Challenges and Methodologies*, Springer, 2005.
6. Van de Goor, *Testing Semiconductor Memories - Theory and Practice*, ComTex Publishing, 1998.
7. T.Powell, W-T.Cheng, J.Rayhawk, O.Samman, P.Policke and S.Lai, "BIST for Deep Submicron ASIC Memories with High Performance Applications", International Test Conference, 2003, pp. 386-392.
8. C. Kern and M. R. Greenstreet, "Formal Verification in Hardware Design: A Survey", *ACM Transactions on Design Automation of Electronic Systems*, Vol. 4, April 1999, pp.123-193.
9. S. Berezin, S. Campos and E. M. Clarke, "Compositional Reasoning in Model Checking", *Lecture Notes in Computer Science*, Springer-Verlag, Vol. 1536, pp.81-102, 1998.
10. P.H.Bardell, W.H.McAnney and J.Savir, *Built-in Test for VLSI: Pseudorandom Techniques*, John Wiley & Sons, 1987.

… Eighth International Workshop on Microprocessor Test and Verification

On Automatic Test Block Generation for Peripheral Testing in SoCs via Dynamic FSMs Extraction

D. Ravotto, E. Sanchez, M. Schillaci, M. Sonza Reorda, G. Squillero

Dipartimento di Automatica e Informatica, Politecnico di Torino, Italy

{danilo.ravotto, edgar.sanchez, massimiliano.schillaci, matteo.sonzareorda, giovanni.squillero}@polito.it

Abstract— Traditional test generation methodologies for peripheral cores resort heavily to low-level descriptions of the circuit, leading to long generation times. Methodologies based on high-level descriptions can only be used if a clear relationship exists between the measured high-level coverage and the gate-level fault coverage. Even in medium complexity circuits, however, a direct relationship between code coverage metrics and fault coverage is not guaranteed, while other RT-level metrics require an effort comparable to the use of low-level descriptions. To overcome this problem, in the case of peripheral cores, a new approach is proposed: FSMs embedded in the system are identified and dynamically extracted via simulation, while transition coverage is used as a measure of how much the system is exercised. Model extraction and coverage maximization are performed concurrently in a completely automated way. This new technique is exploited to drive an unsupervised methodology for generating tests for peripheral cores. Experimental analysis shows the effectiveness of the approach.

I. INTRODUCTION

The System-on-Chip (SoC) paradigm, introduced in the early 90s, has become the most commonly adopted methodology for integration and reuse of electronic devices.

SoCs can integrate into a single chip one or more processor cores with standard peripheral memory and application-oriented logic modules. This high integration of many components leads to an increased complexity of the test process since it decreases the accessibility of each functional module into the chip. Thus, the increasing use of SoCs is leading to new issues on production testing methodologies.

SoC modules can be tested using both *Hardware-based techniques*, such as scan chain insertion or other built-in self test (BIST) structures [11], and *Software-based Self-test* (SBST), whereby a program is executed on the processor core to extract information about the functioning of the processor or other SoC modules and provide it to the external test equipment [2]. There are several reasons to use the SBST: it allows cheap at-speed testing of the SoC; it is relatively fast and flexible; it has very limited, if any, requirements in terms of additional hardware for the test; it is applicable even when the structure of a core is not known, or

can not be modified. Even though SBST is currently being increasingly employed, the real challenge of software-based testing techniques is to generate effective test programs.

SBST techniques have been exploited mainly for the test of microprocessor cores; traditional methodologies resort to functional approaches based on exciting specific functions and resources of the processor [1]. New techniques, instead, differ on the basis of the kind of description they start from: in some cases only the information coming from the processor functional descriptions are required [3]; other simulation-based approach require a pre-synthesis RT-level description [4] or the gate-level description [5].

Simulation-based are heavily time consuming strategies, thus the use of RT-level descriptions to drive the generation of test sets is necessary to allows much faster evaluation. Relying on high-level models not only helps the user of the SoC to perform more simulations increasing the confidence in the generated tests, but is also of value to the manufacturer allowing early generation of a significant part of the final test set. Whereas the correlation between RT-level code coverage metrics (CCM) and gate-level fault coverage is not guaranteed in the general case, several RT-level based methodologies maximize the CCMs to obtain a good degree of confidence on the quality of the generated test set.

This paper proposes the exploitation of an RT-level metric based on an automatically and dynamically constructed FSM. The proposed approach is used to automatically generate test sets for different type of peripheral cores embedded in a SoC. Experiments show that the introduction of the FSM transition metric leads to higher fault coverage and better correlation with gate-level stuck-at fault coverage.

The use of FSM-based coverage metrics has been undertaken in the past [14] [15]. Previous approaches rely on a statically extracted FSM model of the circuit, and require a deep knowledge of the circuit architecture and behavior, and a complex extraction procedure performed manually by the designer.

On the contrary in this work the FSM is dynamically explored using only RT-level information, and no prior knowledge about the interactions between modules is required.

The generation process is driven by the transition coverage on the peripheral's FSM and by the RT-level Code Coverage Metrics (CCMs). Exploiting the correlation

*Contact Author: Giovanni Squillero, Dip. Automatica e Infomatica, Politecnico di Torino, C.so Duca degli Abruzzi 24, 110129, Torino ITALY, Tel: +39 011 090 7186, Fax: +39 011 090 7099, Email: giovannisquillero@polito.it

1550-4093/08 $25.00 © 2008 IEEE
DOI 10.1109/MTV.2007.14

between high-level and low-level metrics, during the generation process only logic simulation is performed allowing the reduction of the generation time. The results are finally validated running a gate-level fault simulation.

Results show that the combination of the FSM's transition coverage and CCM can effectively guide the test block generation performed on RT-level descriptions and that in this way a high fault coverage can be achieved. Moreover, we show that the new approach makes the test generation process more robust, reducing the deviation between the results produced by different experiments and improving the relationship between high- and low-level metrics.

The rest of the paper is organized as follows: section 2 recalls some background concepts, describes the RT-level coverage metrics suitable for test generation, and introduces some considerations regarding the generation process; section 3 outlines the methodology adopted for the generation of test sets for peripheral testing. Section 4 introduces the experimental setup, describing the case study and present the experimental results. Finally, section 5 draws some conclusions.

II. PERIPHERAL TESTING

A. Basics

Generally speaking, a SoC is composed of a microprocessor core, some peripheral components, a few memory modules, and possibly customized cores. Figure 1 shows the test setup postulated for the application of the SBST methodology. An external ATE is supposed to be available for test application: its purpose is to load a test program in the microprocessor memory, start execution, and interact with the peripherals applying data to the input ports and collecting values from the outputs while the program is running.

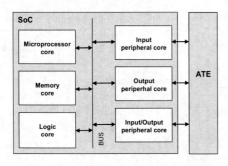

Figure 1. Block diagram of a SoC

To make effective use of the test setup both the test programs and the peripheral input/output data have to be specified; therefore, a complete set for testing peripheral cores is composed of some *test blocks* [9], defined as basic test units composed of two parts: a configuration and a functional part. The configuration part includes a program fragment that defines the configuration modes used by the peripheral, and the functional part contains one or more program fragments that exercise the peripheral functionalities as well as the data set or stimuli set provided/read by the external test machine. Figure 2 is a conceptual view of a test block.

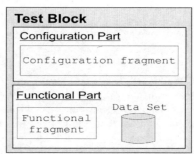

Figure 2. Conceptual view of a test block

Researchers have long sought high-level methodologies to generate high quality test sets. However, despite the efforts, the correlation between high-level and low-level metrics still remains vague in the general case, especially when large combinational blocks exist in the design, whose testability can barely be forecast when resorting to high-level metrics only. While inaccurate to get a quantitative measure, high-level metrics have been fruitfully used as proxies to drive test-set generation. Differently from processor cores, in the case of peripheral cores, large combinational blocks are usually missing, so there is a correlation between high-level metrics and fault coverage. This is not complete but, as experimentally showed in [10], suitable for test set generation.

Therefore, an automatic methodology for the generation of test sets for peripheral cores that uses a high-level model of the peripheral in the generation phase is an interesting solution to overcome new testing issues on SoCs.

As mentioned in [9], traditional code coverage metrics suitable for guiding the development of the test sets for peripheral cores are: *Statement coverage (SC), Branch coverage (BC), Condition coverage (CC), Expression coverage (EC), Toggle coverage (TC)*. Maximizing all the coverage metrics allows to better exercise the peripheral core. Many authors hold that it is not possible to accept a single coverage metric as the most reliable and complete one [6]; thus, a coverage of 100% on any particular metric can hardly guarantee a 100% coverage on the stuck-at faults. Nowadays, thanks to modern logic simulators features, different metrics can be exploited to guarantee better performance of the test sets [7]. Therefore, the test set generation trend is to combine multiple coverage metrics together to obtain better results. It is essential to carefully choose the set of metrics to be maximized to reduce redundant efforts on the test set generation.

B. Previous Work

An attempt to provide effective solutions for peripheral test set generation is presented in [9]; the process is addressed by hand and mainly relies on the experience of a test engineer, who must have a deep knowledge of the peripheral high level description. The goal of the test engineer is to produce a set of test blocks in order to maximize the CCMs. The order in which the CCMs will be maximized must also be selected. In this work every test block follows a quite rigid framework that allows an easy configuration of the peripheral cores in all possible operation modes by only changing a few parameters; and additionally, to set up a carefully chosen stimuli-set based on the specific test block goals.

The process goes as follows: in the first step, an initial test block is generated, based only on functional information about the targeted peripheral, to maximize the statement coverage; then, the generated test block is simulated using a RT-level description, gathering the first code coverage metrics figures. Based on the obtained information, additional test blocks are generated, with the goal of maximizing the rest of the chosen metrics.

Once the first coverage metric is saturated, another one is tackled in order to increase the testing capabilities of the final test set [7]. This process is repeated until sufficiently high coverage values are obtained for all the chosen metrics. Remarkably, it must be taken into consideration that for different peripherals the metrics chosen as critical may also differ, as well as the number of considered CCMs for sufficient code coverage.

In [8] a pseudo-exhaustive approach to generate functional programs for peripheral testing was presented. The proposed method generates a functional program for each possible operation mode of the peripheral core in order to generate control sequences which would place the peripheral in all possible functional modes. The method exploits the capacity of the embedded microprocessor to test peripheral cores at-speed. The pseudo-exhaustive approach was employed to test an *Intel 8251 Universal Synchronous Asynchronous Receiver Transmitter* peripheral core and about 68% of fault coverage was obtained.

The pseudo-exhaustive approach presented in [8] produces a large number of functional programs, since one has to be written for every operation mode; this implies a large application time and a considerable memory occupation, making the method unsuitable in several cases, such as for on-line application.

In [16] the authors describe a generic and systematic flow of SBST application on two communication peripheral core: *Universal Asynchronous Receiver Transmitter and Ethernet controller*. The methodology needs a different test development strategy for the different functional block in the SoC and deep knowledge of the peripheral core is necessary.

In [10] the peripheral test set generation has been automated using an evolutionary algorithm, called μGP^3. The test block generation was supported by the construction of couples of templates: one for program and the other for data generation. The evolutionary algorithm is used to optimize parameter values, leaving the structure of the test block fixed.

The obtained results compare favorably with respect to the manually generated [9], but the setup time required to start the experiments is very high because it involves a deep knowledge of the peripheral core to adequately assemble the test blocks templates.

In [13] an improved version of the evolutionary algorithm has been described, able to optimize both the structure and the parameters. The same results as [10] are obtained with no need of the rigid templates used previously, reducing significantly the required generation time.

III. PROPOSED APPROACH

A. FSM extraction

As stated before, traditional CCMs extracted at the RT-level do not, in general, show a tight correlation with gate-level fault coverage.

One way to model a system is to represent it with a finite state machine (FSM). Coverage of all the possible transitions in the machine ensures thoroughly exercising the system functions.

The proposed approach is based on modeling the entire system as a FSM which is dynamically constructed during the test generation process. So, differently from other approaches, the FSM extraction is fully automated, and requires minimum human effort: the approach only requires the designer to identify the state registers in the RT-level code and does not require any deep knowledge of the peripheral and its hierarchy between modules.

Given the dynamic nature of the FSM construction, it is not possible to assume known the maximum number of reachable states, not to mention the possible transitions. For this reason it is impossible to have a relative value with respect to the entire FSM.

The use of FSM transition coverage has the additional advantage that it makes the interactions between functional modules in the peripheral explicit. The RT-level descriptions use, especially in the case of complex cores, many modules that interact among each other in order to perform the core functionalities. The traditional CCMs do not consider these interactions and only aim at maximizing the coverage metrics in each module. After the synthesis process, at the gate level, the distinction between modules of a core is less clear and therefore it is important to consider the interactions to enforce a correlation between high-level metrics and low level ones.

As experimentally demonstrated [6], maximizing more than one metric usually leads to better quality tests. Moreover, the FSM metric requires an initial *kickstart* for discovering new states. Thus, we make use of all the available CCMs together.

B. Evolutionary algorithm

For the test-set generation process a problem-independent evolutionary tool called μGP³ has been used. μGP³ uses a set of constraints to describe the syntactic appearance of the solutions, i.e. the fact that the structure of each test block matches the one depicted in sec. II.A, thus limiting the variety of the generated test blocks. The constraints are specified in XML format.

In this case the constraints define three sections: a program configuration part, a program execution part and a data part. The first two are composed of assembly code, the third is written as part of a VHDL testbench. Though syntactically different, the three parts are interdependent.

Figure 3 sketches the proposed methodology.

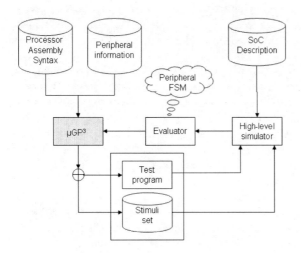

Figure 3. Evolutionary generation loop.

The evolutionary approach generates test blocks starting from information about the peripheral core and the processor assembly syntax only. Every new test block generated is evaluated using a high-level simulator. The evaluation stage assigns a *fitness* to every individual, that is a measure of how well it solves the given problem.

The procedure ends when a time limit is elapsed or when a *steady state* is detected, that is, a predefined number of test blocks are generated without any improvement of the coverage metrics. At the end of the evolutionary run a single test block is provided as output.

The sketched procedure is iteratively repeated to generate a complete test set. In the steps following the first one the evaluation phase is modified in order to only take into account the additional coverage provided by the new test blocks. The rationale for this methodology is that in general it is not possible to completely solve the problem with one single test block. The end result of the process is a set of test blocks that cumulatively maximize the target coverage metrics.

The implemented evaluator collects the output of the simulation and dynamically explores the FSM; it assesses the quality of the test block considering the transition coverage on the FSM and the CCMs.

The fitness is composed of many parts: the FSM transition coverage followed by all the other CCMs (SC, BC, CC, EC, TC). Metrics are considered in order of importance: for example, if two test blocks have the same value of the first metric, the one with higher value of the second metric is preferred, and so on. In this way it is possible, during the generation process, to select more thoroughly those test blocks that are able to better excite the peripheral. The evaluator feeds back the evolutionary core with an indication of the goodness of the newly generated programs.

IV. EXPERIMENTAL ANALYSIS

A. Test case

The benchmark, is a purposely designed SoC which includes a Motorola 6809 microprocessor, a serial communication peripheral (*Universal Asynchronous Receive and Transmit*, UART), a parallel communication peripheral (*Peripheral Interface Adapter*, PIA), a video display unit controller (*Video display unit,* VDU) and a RAM memory core. The system derives from one available on an open source site [12]. The methodology is used to test the UART, the PIA and the VDU in the targeted SoC.

The peripheral are represented as RT-level VHDL code and are composed by different modules. All the core has been synthesized using a generic home-developed library.

TABLE I. IMPLEMENTATION CHARACTERISTICS

DESCRIPTION	MEASURE	PIA	VDU	UART
RT-level	statements	149	153	383
	branches	134	66	182
	condition	75	24	73
	expression	0	9	54
	toggle	77	199	203
Gate level	gates	1,016	1,321	2,247
	faults	1,938	2334	4,054

Table I shows details of the each targeted peripheral, including information at high and low-level. Rows labeled with RT-level present CCM information, such as the number of statements of the peripherals descriptions. The remaining rows illustrate the number of gates counted on the synthesized devices and the number of collapsed faults for the stuck-at model, respectively.

The PIA and the UART can be configured to work in two functional modes with different communication schemes.

The PIA can be used with the following characteristics: polling or interrupt mode; parallel data communication (transmit and receive) with different control schemes.

In the case of the UART the following configuration modes can be used: polling or interrupt mode; serial data communication (transmit and receive) with different data bit numbers, with or without parity, and with 1 or 2 stop bits; serial transmit and receive using different communication rate ratios.

The VDU is a text based display configured with the following characteristic: 80 character across by 25 character down; text buffer memory of 2K bytes; character attribute memory of 2K bytes.

For each peripheral a different test set is generated for better exciting the different peripheral's functionality.

As mentioned before the FSM is dynamically constructed starting from the RT-level simulation. RT-level code must be instrumented to make observable the value of all state registers. The evaluator of the evolutionary tool collects this and filter it in order to identify the global state of the peripheral; every global state in the peripheral represents a possible configuration of values of all the state registers. So, whenever, in the simulation, a state register in any module changes its value, the global state of the peripheral is affected; every newly discovered transition between the global FSM's state is stored in a hash table, for efficiency . At the end of the simulation the number of different keys in the hash table represents the number of transitions. This is fed back to the evolutionary tool together with the traditional CCM's figures, also available from the RT-level simulation.

The number of possible transition could be very high, but this does not impair the methodology. The FSM extraction is dynamic and unsupervised, and only effectively explored transition are taken into account.

At the end of the generation process, some gate-level fault simulation were performed only to validate the proposed methodology; the gate-level fault coverage figures reported in the following sections target the single stuck-at fault model.

B. Experimental results

All the reported experiments have been performed on a PC with an Athlon XP3000 processor, 1GB of RAM, running Linux.

The algorithm parameters for the evolutionary experiments are the same both when targeting only the CCMs, and when the number of transitions in the FSM is also taken into account: μGP³ keeps in every stage a pool of 30 test blocks and generates about 40 of them for every evolutionary step. The description of the syntax for the testbench and the assembly counts about 150 lines of XML.

In order to provide the reader with a reference value, we recall that the fault coverage obtained by the manual approach presented in [9] is 80.96% for the UART and 89.78% for the PIA..

Table II summary the results obtained for the targeted peripherals, reporting the number of FSM transition covered, the high-level CCM's and the stuck-at fault coverage (FC) in percentage. The reader should note that we decided to express the value of traditional CCMs as absolute values (instead of percentages).

TABLE II. RESULTS FOR PERIPHERALS CONSIDERED

	PIA	VDU	UART
FSM TRANSITION	115	191,022	142
STATEMENT	149	153	383
BRANCH	129	66	180
CONDITION	68	23	72
EXPRESSION	0	9	51
TOGGLE	77	191	203
FC(%)	91.4	90.8	91.28

For every peripheral considered the methodology is able to reach a good value of gate-level fault coverage. In the case of the VDU the number of transition is very high; this is due to the state registers that hold the current position in the screen. But since the FSM extraction is dynamic this does not leads to any additional problems.

To demonstrate that the use of the FSM's transition coverage is essential to strengthen the correlation between high an low level metrics 100 experiments on the UART are performed, using both the evolutionary approach presented in [13] and the generation process detailed above.

TABLE III. RESULTS FOR TRADITIONAL APPROACH

	MAX	MIN	AVERAGE	STD. DEV.
STATEMENT	383	381	381.8	0.36
BRANCH	180	178	178.7	0.39
CONDITION	72	70	70.7	0.30
EXPRESSION	51	50	50.7	0.32
TOGGLE	203	201	201.3	0.40
FC(%)	90.7	77.0	84.8	6.37

Table III reports the results of the experiments performed for more thoroughly asses the evolutionary core presented in [13]; the table illustrates the average and the standard deviation of the different coverage metrics (*SC, BC, CC, EC, TC, FC*) and of the stuck-at fault coverage (*FC*). It is worth nothing that the CCMs are very near to the absolute maximum with a little standard deviation while the fault coverage has a greater standard deviation. Even if the values of the CCMs are about the same among all the experiments, the standard deviation in the fault coverage of each test set is relatively high. A large standard deviation implies that the distribution of the solutions found is sparse so there is a considerable difference in fault coverage between the test set found: the methodology, although it obtains good results, is not as robust as desirable, and the obtained solution may not exhibit the expected quality.

This means that the correlation between high-level metrics and low level one should be strengthened.

TABLE IV RESULTS EXPLOITING NEW APPROACH

	MAX	MIN	AVERAGE	STD. DEV.
FSM TRANSITION	142	138	141.0	1.49
STATEMENT	383	382	382.2	0.28
BRANCH	180	178	179.3	0.33
CONDITION	72	71	71.8	0.22
EXPRESSION	51	50	50.8	0.24
TOGGLE	203	201	202.2	0.36
FC(%)	91.3	89.6	90.9	1.10

Table IV shows the results of the experiments performed using also the transition coverage on the peripheral's FSM as a high-level metrics; the table reports the average and the standard deviation of the different coverage metrics (*SC, BC, CC, EC, TC, FC*) and the stuck-at fault coverage (*FC*) as in table III in order to make a fair comparison.

It could be noted that the CCMs are slightly but consistently better than those reported in table III and the average fault coverage is increased by more than 6%. Even more importantly, the standard deviation of the fault coverage is dramatically reduced. This clearly shows that the robustness of the methodology is increased, and solutions of consistent quality can be obtained.

Therefore, the methodology is able to find test blocks that, on average, reach nearly the maximum possible value of CCMs at RT-level and, above all, have a similar fault coverage at the gate level. This experimentally demonstrates that the correlation between high-level metrics and low-level one has been dramatically improved. For this reason the methodology is able to rapidly and effectively generate a test set for the peripheral core.

Table V synthetically reports a comparison between the two methodologies in the case of the UART, highlighting the obtained fault coverage (FC) in percentage, the average generation time (T$_{GEN}$) expressed in hours, the average application time (T$_{APP}$) in clock cycles, and the average size of the test sets, reported as program bytes and data bytes.

TABLE V OVERALL COMPARISON

	FC	T$_{GEN}$	T$_{APP}$	SIZE
[13]	90.7	5.1	28,842	1,953/72
NEW METHODOLOGY	91.3	2.2	32,762	2,345/87

The results clearly show that the new methodology outperforms the previous one in terms of fault coverage and generation time. The latter, in particular, is less than a half with respect to the previous methodology, highlighting the efficiency of the new approach.

Other approaches [8] [16] to peripheral test are not directly comparable with our methodology since they are referred to different devices, although their complexity and the results are similar to the devices analyzed here. Furthermore our methodology only needs RT-level simulation and does not need the time expansive fault-simulation.

V. CONCLUSIONS

This paper presented a new approach for the generation of sets of test blocks for different type of peripheral modules in SoCs driving by the FSM transition coverage and the high-level CCM.

The methodology allows generating test blocks where the relation between high-level coverage metrics and low level one is much stronger; this higher relation has been experimentally demonstrated with a statistical analysis where many test block are generated and evaluated.

The experimental results on different type of peripheral, shown the effectiveness of the proposed methodology.

ACKNOWLEDGMENT

The authors thank Alessandro Aimo, Luca Motta and Alessandro Salomone for their help in designing and implementing the μGP^3, and Alberto Cerato for performing most of the experiments.

REFERENCES

[1] S.Thatte, J.Abraham, "Test Generation for Microprocessors", IEEE Transactions on Computers, Vol. C-29, pp 429-441, 1980.

[2] N. Kranitis, A. Paschalis, D. Gizopoulos, G. Xenoulis, "Software-based self-testing of embedded processors", IEEE Transactions on Computers, Vol 54, issue 4, pp 461 – 475, April 2005.

[3] F. Corno, G. Cumani, M. Sonza Reorda, G. Squillero, "Fully Automatic Test Program Generation for Microprocessor Cores", DATE2003: IEEE Design, Automation & Test in Europe, 2003, pp. 1006-1011.

[4] A. Cheng, A. Parashkevov, and C.C. Lim, "A Software Test Program Generator for Verifying System-on-Chip", 10th IEEE International High Level Design Validation and Test Workshop 2005 (HLDVT'05), 2005, pp. 79-86.

[5] F. Corno, G. Cumani, M. Sonza Reorda, G. Squillero, "An RT-level Fault Model with High Gate Level Correlation", HLDVT2000: IEEE International High Level Design Validation and Test Workshop, 2000.

[6] Jimmy Liu Chien-Nan, Chang Chen-Yi, Jou Jing-Yang, Lai Ming-Chih, Juan Hsing-Ming, "A novel approach for functional coverage measurement in HDL Circuits and Systems", ISCAS2000: The 2000 IEEE International Symposium on Circuits and Systems, pp 217-220, 2000.

[7] E. Sanchez, M. Sonza Reorda and G. Squillero, "Test Program Generation From High-level Microprocessor Descriptions", Test and validation of hardware/software systems starting from system-level descriptions, Ed. M. Sonza Reorda, M. Violante, Z. Peng, Springer publisher, 179 p, ISBN: 1-85233-899-7, pp. 83-106, Dec. 2004.

[8] K. Jayaraman, V. M. Vedula and J. A. Abraham, "Native Mode Functional Self-test Generation for System-on-Chip", IEEE International Symposium on Quality Electronic Design (ISQED'02), pp. 280-285, 2002.

[9] E. Sanchez, L. Veiras Bolzani, M. Sonza Reorda, "A Software-Based methodology for the generation of peripheral test sets Based on high-level descriptions", [accepted for publication on] SBCCI07, Symposium on Integrated Circuits, 2007.

[10] L. Bolzani, E. Sanchez, M. Schillaci, G. Squillero, "An Automated Methodology for Cogeneration of Test Blocks for Peripheral Cores", IOLTS07: International On-Line Testing Symposium, 2007.

[11] R. Chandramouli and S. Pateras, "Testing Systems on a Chip," IEEE Spectrum, Nov. 1996, pp. 1081-1093.

[12] http://www.opencores.org/

[13] L. Bolzani, E. Sanchez, M. Schillaci, G. Squillero, "Co-Evolution of Test Programs and Stimuli Vectors for Testing of Embedded Peripheral Cores.", [Accepted for publication at] CEC: Congress on Evolutionary Computation, 2007.

[14] J. Shen, J.A. Abraham , "Verification of Processor Microarchitectures", VLSI Test Symposium, 1999.

[15] Hoskote, Y.V.; Moundanos, D.; Abraham, J.A, "Automatic Extraction of the Control Flow Machine and Application to Evaluating Coverage of Verification Vectors" ICCD1995: IEEE International Conference on Computer Design, Oct 1995, pp. 532-537

[16] A. Apostolakis, M. Psarakis, D. Gizopoulos, A. Paschalis, "A functional Self-Test Approach for Peripheral Cores in Processor-Based SoCs", IOLTS07 : IEEE International On-Line Testing Symposium, 2007

Automotive Microcontroller End-of-Line Test via Software-based Methodologies

W. Di Palma†, D. Ravotto‡, E. Sanchez‡, M. Schillaci‡, M. Sonza Reorda‡, G. Squillero‡

†Magneti Marelli, Torino, Italy
‡Politecnico di Torino, Dipartimento di Automatica e Informatica, Torino, Italy

Abstract

The car market shows a clear trend towards an increasing presence of electronic devices in engine control systems, due to a strong market drive towards high-performance control systems. Empirical evidence shows that a conventional screening approach only targeting system functionality is not enough to reach the desired high-quality targets. In this paper a methodology is presented to generate a set of test programs that are able to perform a stress test on a widely known microcontroller core. The obtained test set is then characterized in terms of functional coverage.

1. Introduction

The car market shows a clear historical trend towards an increasing presence of electronic devices in control systems for the engine and transmission, the so-called *powertrain*. Customers ask for improved vehicle performance, fuel efficiency, reliability and assistance in driving. At the same time national regulations impose to cut back greenhouse gas emissions and drastically reduce other pollutants through the use of exhaust catalytic converter, possibly restricting circulation to non-compliant vehicles. These factors together determine a strong market drive to design high performance powertrain control systems. The automotive industry answers these requirements implementing powertrain modules able to perform a variety of power and logic functions for the vehicle. To achieve these goals the use of embedded systems based on modern microcontrollers is mandatory to get the necessary computational power.

Failures of electronic devices is currently one of the most relevant reasons for breakdowns in the automotive domain. Empirical evidence shows that a conventional test screening approach only targeting overall system functionality in such a complex design is not effective to reach these high quality targets. This means that individual devices have to be tested inside the system before it is deployed.

Digital devices are usually thoroughly tested by the manufacturer at the end of the production line. However, delivery and system manufacturing conditions may be less than ideal, and defects may be introduced in the produced system. Incoming inspection and system testing are therefore mandatory activities to ensure the target quality. The user of a digital device, however, has no access to the gate-level information available to the device manufacturer, and is not able to generate a test using structural methodologies.

The lack of information makes software-based methodologies for test a privileged choice. These techniques have the advantages of not requiring a deep knowledge about the device under test, of being technology-independent and of allowing the use of cheaper test equipment than possible for structural scan-based test. Also, the generated test cannot be evaluated using fault coverage on the actual object, forcing the use of other metrics.

Several different software-based methodologies have been proposed in the past to test processors. The classic approach [1] models the processor as a directed graph whose nodes are the registers and then excites all the possible transfers between those registers. It is theoretically very powerful, but leads to very long test procedures, both in terms of code size and application time. A newer approach [4] consists in executing relatively short routines, each of which targets a specific processor block, obtaining a high test coverage on that module. This approach, however, does not take into account the control part of the processor, testing it only indirectly. Another methodology [3] resorts to the generation of random code, but also requires the insertion of additional hardware in the processor. More recently, a methodology has been presented [6] to effectively test the data path of floating point units. In [2] another method is described to generate test programs aimed at exercising the arithmetic modules of the processor, but again it does not obtain high test coverage on the control parts. An interesting approach resorts to so-called FRITS kernels [5]: these are programs that repeatedly generate and execute small fragments of pseudorandom code. The methodology, however, requires a deep knowledge of the processor under test to optimally tune its parameters. Apparently no single methodology is able to provide high coverage on all of the processor's functional parts, so a combination of different techniques is desirable.

The environmental and electrical conditions to which the device is subjected during end-of-line testing are often different from those of field operation. The system manufacturer can perform the test using conditions similar to those of the real world. In digital devices some defects may be present that only show up in special conditions, but not in normal environments. It therefore makes sense to perform activities that, strictly speaking, pertain to incoming inspection, during system test.

This paper presents a composite methodology for generation of a set of test programs able to perform a stress test on a widely known microcontroller core, used for an engine control system manufactured by Magneti Marelli. One of the main novelties of the methodology is the collection not only of logical results, but also of timing measurements, to ensure proper functioning of the processor core.

2. Methodology

The activities of incoming inspection demand for environmental conditions more similar to those of in field use than end-of-line test. In particular, the production rates for automotive control systems allow longer test sessions to be carried out than is possible at the end of a semiconductor manufacturing line. Also, it is possible to stimulate the circuit with a workload similar to that expected during normal operation.

SBST techniques rely on the computational power of the processor itself to run a test program on it. Established methodologies are based on running a program able to extract information about the logical functioning of the processor and provide it to the tester. Logical information alone, however, may not be enough.

Modern processors employ several different mechanisms to enhance performance or to ensure device reliability. Typical examples are speculation, such as branch prediction, for performance enhancement, and frequency scaling for temperature protection. Speculative modules by definition may go wrong, but this should not affect the outcome of the elaboration. Thus, a fault in a speculative module may not translate in erroneous computation results, but just in decreased performance. Likewise, an excessive current consumption in one of the functional modules of the processor may cause overheating and decrease the overall system reliability. Thermal protection systems may compensate for this at the cost of reducing the elaboration speed. In both cases, a program takes significantly longer than expected for its execution. This erroneous timing points to an underlying defect, and is an essential metric to detect it.

Another point worth noting is that end-of-line microprocessor test must comply with very strict time limits, as every second of ATE utilization costs money. The employment of SBST to test the device allows to use a lower performance, and therefore cheaper, test equipment than necessary in the manufacturing environment. This in turn permits longer times for processor test, making it possible to approach more closely the workload conditions existing on the field than it would be achievable by the manufacturer.

In particular, it is possible to subject the microprocessor to a *stress test*. In the case of microprocessors the stress test can take on several different forms, relating to the variable considered. The application of high temperature and/or humidity leads to a methodology called *burn-in* test: its main purpose is to eliminate parts whose reliability is too low for normal applications. The combined application of different supply voltages and clock periods is used to derive so-called *schmoo plots*, that define an area of logically correct operation for the device. On the other hand, the definition of a given workload is not trivial for microprocessor test, since its exact amount depends on the details of the microarchitecture. Control of a single variable does not allow to meaningfully apply a stress test.

For pipelined microprocessors, it is the sequence of instructions, and not only the single ones, that determines the behavior of the machine. The number of different instruction sequences is therefore a meaningful indicator of the usage level of the processor's pipeline. One hallmark of a stress test is that the processor is exercised for as long a period of time as possible.

A set of programs that implements a stress test is composed of two main subsets: *focused* and *non-focused* programs. Focused programs are those made to specifically stimulate one module of the architecture. They include programs that test the functional units of the processor, those that saturate internal resources such as a multiplier, and those that exhaustively apply all possible opcodes and addressing modes. The purpose of non-focused programs, instead, is to globally exercise the processor, making it execute very diverse code.

On the basis of the above points, the proposed methodology is composed of a mix of established SBST techniques. The first step tests the register transfers, based on the classic methodology described in [1]. Subsequently, the arithmetic and logical functional units are targeted, using an approach similar to [4]. Then the floating point unit is exercised with a methodology similar to [6]. Another test is performed on the floating point unit using the patterns from the classic approach for integer units [4]. The pipeline is then targeted using a program manually developed by a test engineer. This program contains two main sections: one in which instructions are executed that have no data dependency from the previous ones, and one in which the dependency exists. This second part is further subdivided in several sections. In the first the dependent instructions are consecutive, in the others each couple of dependent instructions is separated by one or more independent instructions. The purpose of this methodology is to verify that all the forwarding mechanisms nominally available are actually active, and also that they are not used when there is no data dependency. After that, a test is performed on the branch unit: the processor state is purposely set, then a series of basic blocks is executed. These basic blocks are composed by a series of data processing instructions whose purpose is to compute a signature for the test, ended by a single branch. To actually make the branches affect the sequential flow of the instructions the basic blocks are executed in a pseudorandom order. All this code is in turn executed inside a large loop. The initial processor state is chosen so that all branches are taken. This approach does not still ensure that the whole code does not result in an infinite loop. To enforce termination of the algorithm the branch condition codes are divided in classes, for instance the "greater or equal" class and the "less than or equal" class. Every class is associated with a register providing the state for the class. Before every branch, the state is updated in such a way that the number of taken branches for every class cannot increase. More formally, if a bit is assigned to every branch, where '1' means that a branch is taken and '0' means it is not taken, and the complete bit sequence is considered, then the processor state is updated so that the corresponding number starts as all ones and decreases monotonically over time. When the bit string is composed of all zeros, the basic blocks are executed sequentially. The state of the external loop is distinct from that of the branch classes. Finally, the pseudorandom test phase is undertaken. The program set incrementally computes an arithmetic signature composed by the content of all the registers. The global test information is composed by this signature and a measure of the time needed to execute the code.

3. Case study

The methodology described above is applied to an engine control system for a direct injection diesel engine (*MultiJet™*), compliant with Euro 5 regulation, and manufactured by Magneti Marelli. To match the computational needs the Freescale PPC5553 microcontroller is used [7] [8], running at 80MHz. The processor core is a 32bit, single-issue implementation of the PowerPC architecture, including a 7-stage pipeline, one integer unit for single- and multi-cycle operations, a vector unit also performing floating point computation, a branch processing unit with a branch target buffer (BTB), and 8kiB unified cache. The controller also includes 1.5MiB flash memory and 64kiB static RAM. Additionally, both a time base and an externally clocked timing unit (eTPU) are available. The microcontroller implements the user level specification of the PowerPC architecture, extending it with a signal processing module and various interfaces.

The complete system runs a control software that occupies about 70% of the flash memory. Since it is part of an automotive system, the core runs a real-time environment.

The test is to be performed offline, during system manufacturing. As the manufacturing rate is in the range of 1 million units per year the total duration of the test has to be less than 6 minutes to perform all test activities, including subjecting the system to a predefined thermal profile and verifying power output and consumption. The time budget for functional test of the microcontroller is 10 seconds, about 3 of which are necessary to load the test program on the microcontroller and unload the results.

Since the core runs at 80MHz, this leaves about 500 million cycles available for test.

Almost all tests are designed to incrementally compute a signature during execution. This is then provided back to the test machine through a serial connection. As time is a critical measure to determine the test outcome, it is measured in two independent ways, and then this information is added to the signature. The first measure is made using the time base registers, allowing to approximately count the clock cycles elapsed for the test. The second is is read from the eTPU; since this is clocked externally, it allows to measure wall-clock time. Both measures are necessary to reliably ensure correct execution. In fact, the number of elapsed clock cycles indirectly measures the amount of computation performed, whereas actual time allows determining the speed at which this computation has been executed. Both measures should be within an expected range to conclude that the device is working correctly. In summary, a correctly working device has to produce the right logical value through the right amount of computation within the correct time.

To perform the functional test five main sections of code have been implemented. The first one tests the correct working of the general-purpose registers, and is derived from the original methodology detailed in [1]. Second, a series of short routines, taken from [4] and [6], is applied to test the integer and floating point units. Then a code fragment implements the test of the forwarding mechanisms. Since the execution module of the pipeline is implemented in three stages then the data dependent part of the code is split in three subsections: in the first the data dependent instructions are consecutive; in the second each couple is separated by one independent instruction; in the third they are separated by two independent instructions. The approach has been implemented using instructions with different timings: single cycle instructions and three-cycle instructions. Actually the pipeline also contains a divider stage, but that is not pipelined, so it has been neglected. Subsequently a section of code containing a subset of all possible branches is executed. As described before, this section is composed of a series of basic blocks executed in a pseudorandom order. The initial state of the processor is chosen so that all branches are taken at least once. Not all possible branches are used for practical reasons: absolute branches can only have as target instructions within a small address range; the processor implements very complex and seldom used branch mechanisms, such as decrementing a count register, checking its equality with zero, branching to the address contained in the link register and saving the return address in the same link register. The use of such complex instructions is clearly limited. These four are focused code sections. Finally, a non focused section is run. This is obtained using an instruction randomizer that obtains information about the processor ISA through an external library. In this way the user can target the test generation to different microprocessor cores, possibly from different families, and can also easily decide which opcodes to use during the non-focused test and which to avoid. This is useful because many microprocessors provide instructions that require special care to use, such as cache management or multi-processor synchronization. These instructions could hardly be expected to work at all inside random code, and are better handled in hand-written tests.

The randomizer used allows to make sure all opcodes in the library are used at least n times. The size of the code is predetermined to fit a memory budget, and the code generated by the instruction randomizer may not fill it. The remaining space is filled by pseudo-random code. This code can be seen as both focused and non-focused. Indeed it extensively exercises the instruction decoder, implementing a nearly exhaustive test, but it also generally excites the processor's functions. The first four code sections have been written by hand, whereas the rest is automatically generated.

For the production test it has been decided to limit the occupied memory by code and data to 64kiB, but statistical results have been collected also for other allocated sizes.

3.1. Test characterization

Since no structural description of the microcontroller is available the results cannot be expressed in terms of fault coverage, but only in terms of functional coverage.

The focused programs are characterized by the patterns they provide to a specific functional unit, and the expected fault coverage for that unit. Both the register test procedure and the short routines targeting the arithmetic unit are known to achieve a high fault coverage and very high functional coverage on their targets [4]. The test program targeting the forwarding mechanism is designed to cover all possible forwarding paths, considering the fact that only three, out of the seven pipeline stages, can provide data to a subsequent instruction. Likewise, the branch testing routine is designed so that every branch instruction is executed at least two times, at least once as a taken branch and at least once as not taken. Actually, the majority of branch instructions are taken several times before switching to non-taken. Thus the targets for these instructions are stored in the branch target buffer (BTB). The storage of the branch address in the BTB and its subsequent removal when the branch is no more taken also affect the total execution time. The target unit for an instruction randomizer is the instruction decode unit. Since all opcodes are used this amounts to a thorough testing. At the same time, executing the opcodes in a random order makes it possible to use the randomized opcodes as non-focused code.

In table 1 the application times for the five test sections are reported. It can be noticed that the non-focused code takes much longer (about three orders of magnitude) than focused code.

As described above, not all instructions in the ISA have been used for non-focused code.

Table 1: Application times for the test sections

Target module	Execution time
Register	91.20µs
Datapath (ALU + FP)	899.02µs
Pipeline	39.03µs
Branch unit	49.95µs
Non-focused	2.994s

The instructions have been classified into several categories: Book E (those that a processor core must provide to be a compliant PowerPC architecture), cache locking, debug, Book E implementation standard (EIS), signal processing engine (SPE), scalar SPE floating point (SPFP), vector SPFP, variable length encoding (VLE) 16 bit, VLE 32 bit. Not all types of instructions are used during operation of the system, so they have not been included in the library. The excluded categories are SPE, vector SPFP and both VLE. This is important to note, because the excluded categories amount to a large number of opcodes.

Of the remaining opcodes those belonging to the cache locking, debug and EIS classes have not been used in the implemented tests, as they require specific programming sequences. Table 2 reports the coverage figures for the five instruction classes considered.

Table 2: Opcodes covered for the considered instruction classes

Inst. class	Used	Unused	Total	Coverage
Book E	150	54	204	73.53%
Cache l.	0	5	5	0.00%
Debug	0	1	1	0.00%
EIS	0	1	1	0.00%
Sc. SPFP	23	0	23	100.00%
Total	173	61	234	73.93%

It can be noticed that not all the Book E opcodes have been used. These can be further divided in several subclasses to better gauge the properties of the non-focused test.

Table 3: Covered opcodes for subclasses of the Book E instructions

Subclass	Used	Unused	Total	Coverage
Branch	1	11	12	8.33%
Cond. R.	11	0	11	100,00%
Inst. Syn.	0	1	1	0.00%
Integer	138	13	151	91.39%
PCRM	0	8	8	0.00%
St. Ctrl.	0	16	16	0.00%
Sys. Lin.	0	3	3	0.00%
Other	0	2	2	0.00%
Total	150	54	204	73.53%

Table 3 reports the details of the used instructions. In

the first column Cond. R. denote instructions that operate on the condition registers, Inst. Syn. are those that perform instruction synchronization, PCRM stands for process control register manipulation, meaning instructions that operate on registers for process control, St. Ctrl. are instructions for storage control, and Sys. Lin. stands for system linkage. The greatest part of unused opcodes is clearly critical when used in random code. Branch instructions are hardly used, but this is not a problem because a focused test exist for the branch unit.

The purpose of the non-focused code is to cover as many working cases as possible on the processor core. For this reason it is meaningful to measure the number of different instruction sequences executed by the core. The PPC5553 features a seven-stage pipeline and is a single-issue machine, so instruction sequences are significant up to a maximum of seven instructions. In the following the coverage figures refer to the used opcodes, not to all the available ones.

The available memory space allows enforcing many times the execution of all opcodes. For instance, with 16kiB available the opcodes can be executed from 1 to 16 times before exceeding the memory budget. Varying the number n of executions changes the results in terms of coverage of instruction sequences. Considering the instruction couples, for small code sizes there is a definite coverage maximum for an intermediate value of n, whereas for larger code sizes the maximum coverage of instruction couples occurs for the largest possible value of n. For the longest instruction sequences the opposite behavior can be seen: the maximum coverage occurs for the minimum value of n, that is 1. This different behavior is directly linked to the fraction of sequence space covered by the code: for a small coverage it is better to fill the space in a random way, whereas for larger coverages it is possible to increase the coverage by enforcing additional bounds.

Since the coverages are very low for sequences of more than four instructions, it is interesting to consider the maximum coverage that can possibly be obtained with every memory budget for instruction couples and triples.

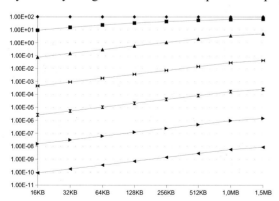

Figure 1: Coverage optimizing instruction couples

The various lines represent the coverages for single instructions (top, guaranteed to be 100%), couples, triples and so on, up to sequences of seven instructions.

The figures are low for the longest instruction sequences because the number of possible sequences is very large, on the order of 10^{18}. The reported data are consistent with what can be expected following a statistical analysis.

It can be noticed that the coverage for instruction couples is near 100%, and also that a saturation effect exists. Increasing the memory budget does not pay off for the coverage of couples as it does for other sequences.

The generation times for the non-focused code scale about linearly with memory size and tend to increase for large values of n.

Table 4: Generation times for non-focused code

Code size	Time (n=1)	Max n	Time (n=max)
16kiB	5s	16	4s
32kiB	9s	32	7s
64kiB	18s	65	17s
128kiB	40s	130	44s
256kiB	97s	260	124s
512kiB	230s	520	412s
1MiB	621s	1040	1477s
1.5MiB	1167s	1560	3097s

The total coverage figures for the complete test set has been computed taking into consideration not only the focused and the non-focused code, but also the initialization sequences necessary for correct execution of the programs. The overall figures are reported in the tables below. The final coverages are unevenly distributed among the considered subclasses: the scalar signal processing and floating point instructions are completely covered, whereas the instructions belonging to the Book E are partially covered. The instructions belonging to the other considered classes are not used at all.

Table 5: Total coverages for the considered instruction classes

Inst. class	Used	Unused	Total	Coverage
Book E	173	31	204	84.80%
Cache l.	0	5	5	0.00%
Debug	0	1	1	0.00%
EIS	0	1	1	0.00%
Sc. SPFP	23	0	23	100.00%
Total	196	38	234	83.76%

Again, a subdivision of the Book E opcodes in subclasses allows to better understand the characteristics and limitations of the methodology.

Branch instructions are used except for the absolute branches. This has a simple reason: the address field for absolute branches is a 16 bit wide signed integer, so they can only have as target an instruction in the first 32kiB of memory; the test routines, however, are not guaranteed to reside in a specific memory section. This forces the exclusive use of relative branches. The missing integer instructions deserve an explanation: actually, of the nine missing instructions two are traps, four are load/store instructions with reverse endianness, two are synchronized memory reads and writes, and one is an instruction that copies the exception register in the condition register.

Table 6: Total coverages for the subclasses of Book E instructions

Subclass	Used	Unused	Total	Coverage
Branch	8	4	12	66.67%
Cond. R.	11	0	11	100,00%
Inst. Syn.	1	0	1	100.00%
Integer	142	9	151	94.04%
PCRM	4	4	8	50.00%
St. Ctrl.	4	12	16	25.00%
Sys. Lin.	1	2	3	33.33%
Other	2	0	2	100.00%
Total	173	31	204	84.80%

4. Conclusions

A SBST methodology has been presented, aimed at testing the microcontroller core in an automotive engine control system. This methodology is applied during system test to subject the core to operating conditions similar to those present on the field, although it would be, strictly speaking, an incoming inspection activity.

The processor core is tested with both *focused* code, to specifically exercise target functional modules, and *non-focused* code, to cover cases not explicitly forecast. The test results are collected in the form of a signature, incrementally computed during execution, and two time measurements, one in terms of clock cycles, the other of wall-clock time, to ensure correct functioning of the device.

The characterization of the resulting test clearly shows the applicability and limitations of the methodology.

The methodology can be readily integrated within an industrial environment.

5. Acknowledgment

The authors want to thank Andrea Maffiotto for performing most of the test generation and all the characterization.

References

[1] S. Thatte, J. Abraham, "Test Generation for Microprocessors", IEEE Trans. on Computers, Vol. C-29, June 1980, pp. 429-441
[2] L. Chen, S. Dey, "DEFUSE: A Deterministic Functional Self-Test Methodology for Processors", IEEE VLSI Test Symposium, 2000, pp. 255-262
[3] K. Batcher, C. Papachristou, "Instruction Randomization Self Test For Processor Cores", IEEE VLSI Test Symposium, 1999, pp. 34-40
[4] Paschalis, A.; Gizopoulos, D.; Kranitis, N.; Psarakis, M.; Zorian, Y., "Deterministic software-based self-testing of embedded processor cores", IEEE Design, Automation and Test in Europe 2001, pp. 92-96
[5] Parvathala, P.; Maneparambil, K.; Lindsay, W., "FRITS - a microprocessor functional BIST method" International Test Conference, 2002, 590-598
[6] Xenoulis, G.; Psarakis, M.; Gizopoulos, D.; Paschalis, A., "Testability Analysis and Scalable Test Generation for High-Speed Floating-Point Units", IEEE Transactions on Computers, Vol. 55, Issue 11, Nov. 2006 pp.1449-1457
[7] Freescale Semiconductor, Inc., "e200z6 Reference Manual"
[8] Freescale Semiconductor, Inc., "MPC5554 and MPC5553 Reference Manual"

Functional Validation and ATPG
MTV 2007

Intel® First Ever Converged Core Functional Validation Experience: Methodologies, Challenges, Results and Learning.

Tommy Bojan, Mobility Group, Intel Corporation[1]
Igor Frumkin, Mobility Group, Intel Corporation[2]
Robert Mauri, Digital Enterprise Group, Intel Corporation[3]

ABSTRACT

Intel® Core™ microprocessors, including Xeon® 5100 (codenamed Woodcrest) and Core™ 2 Duo (codenamed Conroe and Merom), was the first Intel® Converged Core product which simultaneously hit all market segments (server, desktop and mobile platforms). A year after these microprocessors have successfully entered the market and significantly improved Intel® revenues and competitive position, it is time to analyze the post silicon validation experience.

This paper discusses the Core processor's validation challenges, among them were:
- Very aggressive delivery schedule. The parallel validation of three market segment products within a one year timeframe (first silicon to Product Revenue Qualification).
- Dramatic difference of Core™ Architecture and micro-architecture from those of Pentium® 4 family microprocessors which held the desktop and server market segments for the prior six years.

The paper describes the post silicon functional validation methodologies, on both the System Validation (SV) and Compatibility Validation (CV) disciplines, at Intel Corporation as well as original equipment manufacturer (OEM) engagements during the validation cycle. The validation strategy was to quickly ramp up the internal validation capabilities and uncover all silicon issues within Intel Corporation. The overall goal was to preserve the tight OEM partner development cycles from samples to launch.

The paper summarizes the major results vs. expectations and key learning for the future products.

1. INTRODUCTION

Intel® Core™ microprocessors, including Xeon® 5100 (codenamed Woodcrest) and Core™ 2 Duo (codenamed Conroe and Merom), was the first Intel Corporation Converged Core product which simultaneously hit all market segments (server, desktop and mobile platforms). Numerous post silicon validation challenges of Core™ 2 Duo microprocessors included:
- Very aggressive delivery schedule of one year timeframe from first silicon to Product Revenue Qualification, which could be achieved only with very minimal number of intermediate steppings. It dictated the steepest ramp in uncovering the silicon component bugs and system compatibility issues.
- The parallel validation of all three market segment (server, mobile, desktop) products. Despite common Intel® Core™ architecture each of them had unique, for specific segment, features and was integrated into different platforms, usage models and software.
- Dramatic difference of Core™ Architecture and micro-architecture from those of Pentium® 4 family microprocessors which held the desktop and server market segments for many years, bringing a reasonable risk of compatibility issues across HW components and software.

At Intel Corporation, the traditional CPU post silicon platform level validation cycle is a puzzle integrating different activities which still run under single validation program umbrella. Among them are:
- Functional (logic) validation which includes system validation (SV), compatibility validation (CV) activities and Original Equipment Manufacturers (OEM) program.
- Circuit validation across CPU IO and Core domains covered by Electrical Validation (EV) and Circuit Marginality Validation (CMV) teams, respectively (not in the scope of given presentation).

SV prime focus is the microprocessor component logic functionality according to Programmer Reference Manual (PRM) [1]. SV run unique tests that typically have nothing in common with any commercial software on specially designed hardware configurations. The SV platforms aim to simulate all commercial platform configurations including the add-in cards options. Similarly, the SV testing aims to simulate all the existing, and yet to be developed, commercial software code combinations. These two factors provide much better controllability on validation coverage, since they allow for the accurate generation of any Intel Architecture scenario permitted by Programmer Reference Manual (PRM) as well as numerous micro-architectural scenarios and events. The SV environment is also very efficient for failure localization, which guarantees fast issue debugging and root-causing.

The CV objective is to ensure that the microprocessor and chipset are functioning correctly in the end user or customer environment configuration. This incorporates the usage of operating systems (OS), commercial applications and

---

[1] **T.B.** Author is with Intel Corporation, Israel Design Center, P.O. Box 1659, Haifa 31015, Israel. Phone: +972-4-865-5502; e-mail: tommy.bojan@intel.com
[2] **I.F.** Author is with Intel Corporation, Israel Design Center, P.O. Box 1659, Haifa 31015, Israel. Phone: +972-4-865-5335; e-mail: igor.frumkin@intel.com
[3] **R.M.** Author is with Intel Corporation, 2800 Center Drive, MS 301, DuPont WA 98327 USA. Phone : +1-253- 371-9415, e-mail: robert.mauri@intel.com

commercially available add in hardware that run on customer reference boards. CV focuses on the microprocessor and chipset integration into existing or new platforms. During this integration process, CV uncovers CPU and chipset component bugs and issues, as a result of running the unique code flows under the OS. This approach poses some significant challenges in terms of coverage (very large end user configuration space) and a high noise environment (issues can come from all levels of the HW and SW stacks).

The OEMs prime goal is to maintain the tight customers development cycle from samples to launch. Their validation process is very similar to that of CV, with a focus on overall system functionality. Early delivery of engineering samples to OEMs is very important part of overall validation program, as it allows uncovering issues related to compatibility to customer systems as well as the residual component issues which become visible due to specific customer usage models, before many millions of new microprocessors are available on a market.

2. SYSTEM VALIDATION METHODOLOGY

The major principles of the System Validation methodology at Intel Corporation, successfully applied several years ago to Pentium® M microprocessor [2] have been enhanced for the next generation CPUs, namely, for Core™ Duo (codenamed, Yonah) and Core™ 2 Duo. Major methodology enhancements are being aimed to address the increasing complexity of microprocessors, e.g. multi-Core configuration on the same die, more and more aggressive power management scheme, performance micro-architectural enhancements etc. To mention, despite dramatically increased CPU complexity, the overall validation cycle timeline, dictated by market, remains practically unchanged (~ 1 year), that also requires an improvement of the validation work effectiveness.

SV content is based on two approaches, random instruction testing (RIT) and direct testing.

1) RIT concept means that CPU correct functionality is being checked by a tremendous amount of discrete tests aimed at covering all possible architectural and micro-architectural scenarios defined by PRM. Each random test (seed), typically runs from reset, emulates some artificial operational system with all its attributes (e.g., descriptor and interrupt tables, exception handlers, paging hierarchy), and also contains several thousand instructions of "user's code", running under this pseudo-Operating System (OS). Both OS attributes and user's code are maximally random, in terms of instruction sets, memory allocation for OS level structures, code and data, caching policy, etc. Besides, random programming of platform residing components (e.g. some chipset features) and special SV agents are also under seed responsibility. Random test generation is being done in a software environment, specially developed at Intel Corporation.

2) Direct testing applications used in SV are implementing either pseudo random algorithms found efficient for the memory coherency and cache coherency coverage areas, or focus tests for automatic validation of specific corner cases which are difficult to implement in random tool. Unlike RIT seeds which are pretty short (~ tens of ms of execution time on 3 GHz microprocessor) and use an architectural simulator to check correct seed results, direct tests are based on self-checking and may run for as long as hours. The latter makes them efficient bug finders in cache coherency domain where hitting the interesting scenarios may require long exercising of internal arrays.

In both above cases, the test generation rules are based on hierarchy of test plans written by SV engineers per specific CPU feature or group of features. Before implementation, these test plans are subject to thorough reviews with experts across architectural, design and validation teams, for all possible architectural and micro-architectural corner cases coverage. The quality of test plans and of their implementation in SV content does result finally in the success of silicon in mass production (e.g., absence of escapee bugs).

SV platforms rich of special validation hooks is an essential part of SV methodology. These hooks give more controllability from SV test to implement required coverage scenarios, thru enhanced visibility over CPU states across busses and internal structures. A good example is controllability over CPU power management exercising where an agent residing on the Processor System Bus (PSB) captures CPU entering a power state and sends some trigger to external card, to issue snoop or interrupt as a wake event. To mention, both PSB agent and external cards are programmed from inside the test with maximally possible randomization of parameters (relevant for RIT). Another important part of on board SV hooks is Joint Test Action Group (JTAG) controller which provides an efficient access and manipulation with CPU internal registers and Design For Testability (DFT) features from inside the test. Block diagram of typical SV platform is shown in Fig.1.

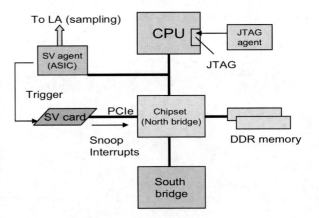

Figure 1: Block diagram of SV platform.

With dramatic increase of the PSB frequency, usage of Logic Analyzer Interface (LAI) as a probing solution becomes problematic and costly. SV platforms design typically includes special ASIC agents ensuring a reliable probing without any impact on signal integrity of the bus.

The tight post silicon time schedule dictates the requirement for SV to uncover the bulk of fatal silicon issues (those which have strong negative impact on CPU delivery schedule or blocking the further validation progress) within the first several weeks. It will ensure their quick root causing and thus finding acceptable workarounds to maximally open the other validation channels (CV and early OEM involvement). To achieve this goal, SV readiness to mass execution, from both content and execution environment perspectives, from the day CPU has booted, is a must. Efficient pre-silicon, aka mock, system level execution is only possible with a reference CPU or several CPUs supporting the maximal possible subset of new microprocessor features. In some cases these reference CPUs are pin or electrically incompatible to the new platform. In such cases special small form factor interposers, equipped with reference processor are being developed. Moreover, they become part of early enabling programs, being delivered to platform development teams inside Intel Corporation (reference boards) and externally to Intel Corporation (OEM systems).

3. CORE™ 2 DUO SV CHALLENGES

Core™ 2 Duo microprocessor design is principally based on Core™ Architecture introduced one year before in mobile Core™ Duo product (codenamed, Yonah). This architecture assumes 2 symmetrical Cores with shared L2 cache and very aggressive power management scheme essential for mobile microprocessors. Expansion of Core™ Architecture to desktop and server segments has resulted in significant design changes among them are 100% increase in PSB frequency, 64-bit Extension Technology (EM64T), new instruction set for multimedia applications (SSE3), enhanced power management capabilities, numerous micro-architectural enhancements aiming to establish leadership in performance benchmarks across all three segments, RAS features essential for servers. Functional compatibility to existing chipsets was also in high validation focus as new microprocessors were replacing Yonah or Pentium® 4 (codenamed, Prescott and Dempsey) as part of respective platforms refresh cycle.

Yonah legacy SV content, which supported the Dual Core architecture and the majority of power management scheme, was taken as baseline, while validation coverage for the new features like SSE3 and EM64T was put upon it. Proper validation of multiple, performance oriented, micro-architectural enhancements was the biggest challenge not just because precise scenarios implementation in SV tools was not trivial but also due to low external visibility of them to happen. Several existing and novel approaches to functional coverage measurements were applied [3] and significantly increased the validation confidence. Coverage profiling in post silicon validation remains an active area of development at Intel.
As new microprocessor was primarily introduced with existing platforms, legacy SV boards were re-used. At the same time some of them, being developed for different processors and SW applications, also by different divisions at Intel Corporation, were short of required SV hooks. To overcome this problem, a special SV hooks card was developed, with generic connectivity to any type of SV board or even to Customer Reference board (CRB).

SV mock execution to ensure the whole SW and HW environment preparedness was done across all types of validation platforms with existing to that time CPUs, namely with Yonah, Prescott or Dempsey (no interposers were needed). As each of these CPUs has supported in turn some unique for new microprocessor features, it allowed to dry run the majority of the SV test content and related on-board hooks. Content for some yet unsupported to that time features was also verified on RTL.

4. COMPATIBILITY VALIDATION METHODOLOGY

The Compatibility Validation (CV) main goal is to validate components (Processor and Chipset) at the platform level by using commercially available operating systems, application software and hardware in a manner that is consistent with the usage by the customer (end user).

From the CPU validation viewpoint, CV testing is designed to validate scenarios that are used in the "real world". This approach differs significantly from the SV approach in that determinism and controllability are difficult to achieve. In fact, the act of imposing these attributes into a CV environment would entail the usage of specialized tools, drivers and other intrusive mechanisms that would affect the systems ability to "naturally" operate.

The overall CV strategy is to apply the testing along the various categories in the form of matrices. The matrix components used for testing Core™ 2 Duo were:
OS load/boot matrix
Memory matrix
IO matrix
Application Software

In addition to the above categories, there are functional feature coverage areas that CV uses.
Power Management
Performance Characterization
Stress/Concurrency
New Features

The CV strategy will be to apply each of these matrix items over the lifecycle of a product to form a coherent end user coverage space. As CV progresses through the validation cycle, the team is faced with three specific challenges:

Challenge 1: Platform stabilization ramp. In projects that include a new platform (new chipset) as opposed to a follow-on processor, the integration process is much more complex. CV teams face an integration challenge with multiple new silicon components, new technologies and new software. The challenge is less severe when the product being introduced is a CPU that will be inserted into a pre-existing platform (already productized). Fig. 2 depicts one component of the complexity. The first part has to do with platform stabilization and capability. At the beginning, the platform lacks the stability and functionality to support the full featured testing as defined by the test plan. The second element is related to test noise. There are many test failures in this early stage and the failures come from the platform and software stack itself. This is all part of the integration phase. The goal of this early phase is to establish stability and to reduce test failure noise (failures due to non component issues).

The CV strategy in this instance is to test and integrate lower level featureset building blocks to establish confidence for the next layer of testing. As a result, a lot of the testing in the early phase is related to basic feature operation, OS load/boots, memory and IO card checkouts (basic functionality established within the OS...drivers loaded etc). Once a reasonable stability metric has been achieved, stress testing starts. Automation hardware is also used in areas like power management testing with a similar stability curve to integrate the infrastructure to the system (support of C and S states, for example).

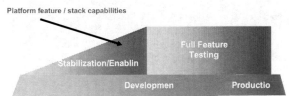

Figure 2: Platform stabilization and testing cycle at CV.

Challenge 2: New feature acceptance or deployment. Some new features require "tall" software implementation stacks. Virtualization, SSE, IO acceleration technology and Speed Step ™ technology are examples of such features. In some cases (Virtualization and Speed Step™), Operating system kernel modifications are necessary. In other cases (IOAT or SSE), applications or drivers are necessary. The issue here is that the software products are usually not time to market (TTM) with the silicon validation cycle, since software and operating system vendors require time for the integration of their software stacks while the system stability prohibits serious integration work in the early phases of the program.

The CV strategy to this particular challenge is to test what is available in the given timeframe and to assess the state of the ecosystem in terms of readiness. This assessment can be characterized in the form of risk to the project (in terms of potential issues in specific silicon feature). The more robust and ready the ecosystem the better testing that occurs prior to the productization/launch phase.

Challenge 3: Given that permutation factors include Operating Systems, system topologies, add in devices and software application stacks, the space that is considered to be valid in an end user configuration is quite large. Also given that CV has a fixed and limited capacity, the team has to draw the line somewhere with respect to selection of test matrices.

The strategy employed for this particular challenge is to leverage the testing of other teams. CV works with OEM's to understand some aspects of their test and test plans in order to gauge the overall coverage of the validation space. This activity is done by working with the customers (through technical marketing channels) to survey their testing (along the CV matrix axes defined above).

To add, the CV team is heavily engaged in customer failure analysis once Intel Corporation component (CPU, specifically for this paper) is prime suspect. This activity has two phases that span a pre-product launch (engagement with OEMs) and a post product launch phase (engagement with a wider array of customers who may experience issues with the CPU). Customer failure reproduction in CV lab is first attempted on Intel system (CRB) but, in many cases, OEM need also to send their specific hardware and software to technical marketing representatives. After failure reproduction, the debug flow starts until root cause and workaround or fix recommendation.

5. CORE™ 2 DUO VALIDATION RESULTS

5.1. System Validation results

SV content and environment readiness has resulted in very steep mass execution ramp up across all three microprocessors (Woodcrest, Conroe and Merom) which started practically the day all CPUs booted. RIT execution with a throughput of several million random seeds a week was also characterized with a very low level of false alarms, so that validation engineers could concentrate on debugging of real issues. Fig.3a shows the cumulative fatal bug detection rate from silicon to launch demonstrating that 70% of them were uncovered within first 6 weeks. Fig. 3b shows the pareto of fatal bugs detected across different validation methods clearly illustrating that RIT methodology accompanied by direct SV testing are the major bug finders, as planned (to note, while the same bug could be exposed in parallel to several tools or methodologies, this graph refers to prime detector only).

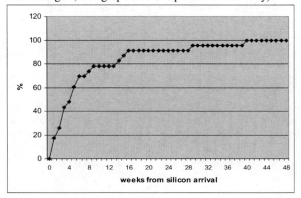

a)

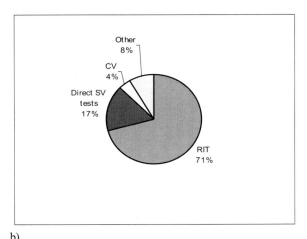

b)
Figure 3: a) Cumulative fatal bug detection rate by Intel Core™ 2 Duo from first silicon to Launch.
b) Pareto of fatal bug detection across methodologies

Another essential part of any microprocessor validation program is the capability to fast and efficient reproduction of the system failures on the RTL model, to allow the design team for root-causing of the failure origin. An automation methodology for RTL failure reproduction (i.e. guaranteeing that CPU behavior on system is exactly reproduced on RTL) has been developed at Intel Corporation and is being applied for years. Nevertheless, as microprocessor complexity increases, the automatic reproduction methodology also faces new challenges and actually runs as a separate, high priority program. It brought the fruits once Core™ 2 Duo post silicon validation started, as reproduction automation has been enabled from early first days. As a result, in the majority of cases, the design teams spent just days for system failure root-causing and, if bug is confirmed, for providing either a workaround (in software or hardware) or clear signature for further screening. Therefore, despite naturally high system failure rate in the first weeks of silicon, the validation teams succeeded to survive and efficiently identify the big portion of critical bugs (as seen on Fig. 3a).

5.2. Compatibility Validation/OEM results

Historically, the CV methodology assumes that the silicon bug and compatibility finding capability tends to be bimodal. This implies that some CPU, platform or software issues are exposed very early with simple functional workloads and very late in the cycle when the more complex workloads are able to run deeper stress cycles. The issues found in the first mode tend to be what is termed "easy" ones, as they can be detected with OS operation and small workloads and tend to be repeatable and often catastrophic. These characteristics allow the issues to be easily identified in a high noise environment. Moreover, the vast majority of real silicon issues related to this mode are also exposed to SV as well (which is more natural environment for debug, as noted above). Instead, the customer issues in this mode are not expected since the CV validation cycle tends to be a leading phase (primarily due to earlier samples).

The data in the second phase or mode also includes issues (both silicon and integration related) reported by OEM's, as silicon starts running through their cycles. In this mode, the failures tend to be much more elusive with lower reproduction rate as they require much more complex sequences and timing of events to occur (a smaller set of workloads and test scenarios are capable of exposing such issues). Issues of this sort can also appear in the early phase but are much more difficult to be diagnosed due to high noise environment. As the platforms get more mature (stable), the failure noise rates decrease over time and this allows the debug teams to have a better ability to discriminate such failures. Debug activity is occurring in the CV validation labs as well as in OEM validation labs simultaneously. This distributed process also introduces latencies as failure data has to be collected across company boundaries and analyzed.

Bimodal nature of CV/OEM cycle is clearly illustrated on Fig. 4 which shows different view point for the Desktop product. The chart depicts the failures reported on a weekly basis. The failures, as shown on graph, clump together around test cycles which can be divided into 4 main groupings. The first three groupings show significant failure spikes while the last grouping shows a smaller increase. This trend is also indicative of the stabilization effect.

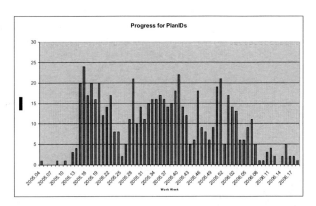

Figure 4: Weekly failure rate chart and raw failure data groupings.

One additional side affect of this bimodal behavior is that the second mode tends to align with the product release window. As a result of the increased stability and the ability to collect data on the failure (this element takes time to accumulate), the OEM's finally have enough data to discern that there is an issue related to silicon. This timing element poses additional pressure on the debug teams in CV to isolate the issue and to compile resolution mechanisms (workarounds or fixes). During the launch cycle of the Core™ 2 Duo product, we did experience a number of these issues. They needed quite a bit of effort to resolve as illustrated in below examples:

USWC speculation issue
The issue deals with a behavior of the CPU with respect to the speculation engine and USWC memory space. Core™ 2 Duo

introduced a very aggressive speculative fetching/reading engine that had different behavior than its predecessors (Pentium® 4). When the issue was first reported by a graphics adapter vendor, it was isolated and believed to be an issue with the graphics adapter driver code that caused accesses into graphics memory prior to its initialization by the driver code. Intel Corporation worked with the vendor to modify their driver code and the issue was thought to be resolved. Much later in the validation cycle (closer to the launch window), it was found that other graphics adapter drivers had similar issues. The debug teams diagnosed this problem further and realized that the issue was not specifically due to the driver code but with the way speculation worked. Even though this issue is not classified as a bug (feature worked as designed), its side affect with the ecosystem was considered to be a compatibility issue (CPU does not work like its predecessor and affects software operation adversely).

Memory Management issue
An issue arose from an operating system vendor (OSV) when they tested an updated version of their operating system. In this particular instance, Intel Corporation had seeded the company with a number of software development vehicles. The early development platforms had similar stabilization issues as all of the other OEM's had experienced. At some point the vendor modified some of the memory management algorithms to increase overall system performance. In a process of debugging their kernel code, they found some sequence of events that led to a data corruption. Further root causing of the failure at Intel Corporation has revealed that the failure scenario utilized a mechanism implicitly not allowed by the Programmer Reference Manual (PRM). What we learned from this experience is that operating systems vendors are using and will use all software means to improve the system performance and that there is an extended gray space in PRM with respect to what is "legal".

6. LEARNING SUMMARY

Overall, a coordinated work of internal system and compatibility validation teams, at Intel Corporation, has resulted in successful delivery of Intel® Core™ microprocessors, including Xeon® 5100 and Core™ 2 Duo, to market in time. Traditional validation methodologies applied by these teams and enriched to meet the new challenges in general have proven themselves.

In line with the validation strategy, the initial validation period, lasting 5-7 weeks, caused a critical positive influence on the whole cycle. During this period, the majority of system level visible bugs were uncovered, mostly by SV, and workarounds were identified to allow full volume enablement of all validation channels. To add, pre-silicon extensive dry run of the whole SV execution/debug environment was key success to the steep post silicon validation ramp up.

The large end user space has been discussed as a challenge for Compatibility Validation, but there is also an aspect of return on investment to consider. As the number of bugs found with the CV methodology is relatively small compared to that of SV, CV should invest more in its strategic partnership with OEM's. A necessity of close work on early enabling of OEM systems and better synch up of validation programs is the key learning here. If quicker enabling and stabilization of OEM platforms is achieved, this would essentially improve the overall validation "system" performance by shifting the ability to discover late arriving bugs and compatibility issues much earlier. To add, this strategy should be accompanied by earlier delivery of silicon engineering samples to OEM's as well as by customer commitment to early validation cycle.

Software vendors invest much effort in performance optimizations operational systems (e.g., memory management subsystem) by using algorithms which fall on gray areas of Programmer Reference Manual (PRM). Therefore, Intel Corporation should focus on proactive investigation into operating system mechanisms on subject of identification of all such cases. The result of this investigation has to be an input to both SV and CV teams which should include these cases in respective validation plans. Early internal validation will allow us to be aware of all system consequences for the given usage models and give trigger to proactive steps, e.g. documenting new guidelines or restrictions in PRM.

REFERENCES
[1] Intel Corporation, "Intel® 64 and IA-32 Architectures Software Developer's Manual" (PRM), Vol. 1, 2a, 2b & 3, Nov. 2006.
[2] Isic Silas, Igor Frumkin, Eilon Hazan, Ehud Mor, Genadiy Zobin "System-Level Validation of the Intel® Centrino™ Mobile Technology-Based Pentium® M Processor." Intel Technology Journal, Volume 8, Issue 2, 2003.
[3] T. Bojan, M. Aguliar, E.Shlomo, T.Shachar "Functional Coverage Measurements and Results in Post Silicon Validation of Core™ 2 Duo Family. Proceedings of 12th Annual IEEE International Workshop on High Level Design Validation and Test. 2007 Irvine, CA.

Eighth International Workshop on Microprocessor Test and Verification

Chico: An On-Chip Hardware Checker for Pipeline Control Logic

Andrew DeOrio, Adam Bauserman, Valeria Bertacco
Dept. of Electrical Engineering and Computer Science
University of Michigan, Ann Arbor
{awdeorio, adambb, valeria}@umich.edu

Abstract

The widening gap between CPU complexity and verification capability is becoming increasingly more salient. It is impossible to completely verify the functionality of a modern microprocessor before shipping, much less before tape out. Recent studies indicate that the majority of errors in these designs are centered on control and forwarding logic[9]. To address this problem, we present Chico, an efficient approach to on-chip hardware correctness that specifically targets escaped design errors in these high risk functional blocks. Our solution includes an on-chip checker block that monitors the correctness of potential data dependencies and program order of the executed instructions before they are allowed to commit. If this online checker detects a mismatch, the processor's exception handler is invoked, reconfiguring the system to a known-correct, formally-verified mode of operation which can correctly re-execute and commit the faulty instruction. The processor can then resume its normal, high throughput mode of operation. In our experimental setup, we have implemented Chico in an out-of-order processor design and evaluated its performance impact on 11 distinct buggy variants of the design running SPECint benchmarks. Our results indicate that Chico can overcome the errors present in these buggy designs at a minimal performance cost, ranging from less than 1% up to 4%. In addition, we evaluated Chico's area cost and found it to be an order of magnitude smaller than other popular solutions such as DIVA[10], with an area impact of less than 3% for our experimental processor. Our approach is novel in that it shows no appreciable performance degradation on a correct design, and it is a low complexity, area-frugal solution compared to previous work.

1. Introduction

The control logic portion of complex digital systems is notoriously difficult to test and debug. The growing complexity of large out-of-order cores further compounds this issue. For these reasons, the correct design of a pipeline's control logic is such a complex problem that it is often unachievable. As a result, this portion of the design is the culprit for more than half of the escaped bugs reported in modern processors' errata documents[8].

Simulation-based verification is the mainstream approach used in industry to validate a design and correct any design errors. However, due to the complexity of the design state space, only a small fraction of the design's functionality can be validated within the available development time window. Even so, bugs are difficult to diagnose and resolve once identified. Frequently, this requires analyzing extremely long simulation traces in order to identify the root cause of a problem. Formal verification techniques, while capable of checking the correctness of a design aspect under all possible execution situations, can only be deployed for very simple designs.

Recently, a few runtime verification solutions have been proposed by the research community [9, 10, 2]. A common trait of these solutions is the use of one or more on-chip checkers which can detect all or some functional errors by monitoring the processor's execution flow. If an error is detected, most solutions will provide a correction mechanism, which enables the processor to overcome the buggy configuration at a performance cost. Hence, the impact of an error in an aspect protected by an online verification mechanism is limited to a graceful performance degradation, rather than incorrect results or, possibly, a system crash. Thus, a key benefit of these runtime verification solutions is that they enable the verification team to focus its efforts on the design's execution scenarios that arise most frequently, and relieve the burden of striving to fully verify a design before its release to the market.

1.1 Contribution of This Work

In this paper we present an on-chip checker solution, called Chico, which specifically targets the most critical control aspects of out-of-order processor designs. Because of Chico's specialized focus, its checking and correction mechanisms require very few hardware resources, resulting in an extremely small area overhead, a full order of magnitude less than some previous approaches such as DIVA[10]. We made the choice to focus exclusively on control logic because of the large fraction of escaped bugs that originate in those blocks, and their critical impact on the system's correct operation.

Chico consists of a checker hardware block that monitors the flow of instructions executed by the processor, and checks that the control aspects of the execution are performed correctly: source value retrieval, data forwarding, branching selection, dynamic execution flow, deadlock, etc. Instructions are checked just before the commit stage, thus avoiding unnecessary expenditure of effort on errors that may appear during speculative operations. If an error is detected, Chico uses the processor's exception handling mechanism to prevent the problematic instruction from completing. It then forces the processor to re-configure into a barebone system by disabling most performance features (including pipelining, data forwarding, branch prediction, etc.) and re-executes the same instruction in this configuration. The barebone version of the design, also called *degraded mode*, is sufficiently simple that it can be formally verified at design time, thus we can guarantee correct completion of the instruction. Once the buggy state has been overcome by degraded mode, the normal mode of operation is resumed along with full-speed execution.

Our solution presents several novel advantages. First of all, Chico requires connections to just a handful of the design's signals, hence it can be easily integrated into a wide range of pipeline architectures. Additionally, our solution is free from false positives, triggering the degraded mode of operation exclusively when a bug arises. In contrast, other solutions described in the literature[9] allow false positives to trade accuracy (and thus performance) for area overhead.

2. Related Work

Runtime verification solutions and the use of on-chip checkers have started to appear only in recent years, one of the first works in this domain being DIVA[10]. DIVA is a solution deploying a simple, but complete, processor core next to a complex core. The simple core re-executes and verifies the results of all instructions about to complete in the complex core. Chico differs from DIVA in that it does not need to recompute results and is therefore able to offer a significant reduction in area cost. Based on our experimental results, Chico's area is an order of magnitude smaller than DIVA. DIVA provides greater coverage than our design, but at the cost of significantly higher hardware overhead and prohibitive wire routing challenges, as it must connect to each stage of the complex core. Moreover, the DIVA checker occasionally causes the fast CPU core to stall, incurring a performance penalty even when no error has occurred. Our proposed method has no performance impact during normal operation, only when a faulty instruction is flagged and re-executed by the checker is any degradation realized. Additionally, the significant extra wiring required by DIVA to connect each processor stage to its checker counterpart is eliminated in the Chico solution. Consequently, Chico is not as susceptible to timing, wiring delay and routing problems. All of the this can be achieved with little sacrifice of error coverage, since control logic is the most frequent source of functional bugs. Finally, our Chico checker can be kept extremely simple because we rely on the main processor itself to recover from an error (by reconfiguring it to a barebone system). In contrast, the DIVA checker includes a complete processor implementation, since the checker is also in charge of the recovery.

Another runtime verification solution is Field Repairable Control Logic (FRCL) [9], a recently proposed method used to correct design faults. Similar to our approach, faulty instructions are executed in a degraded mode to correct errors. However, the error detection method is not as automated as with Chico. Instead, the values of control bits in the processor are matched at runtime against programmed "bug patterns." The method is quite flexible, though bugs need to be manually discovered and characterized by a verification engineer before a pattern can be created to correct them. Because FRCL observes only a small number of signals and has limited storage space for bug patterns, error conditions must often be over-specified. This leads to a high incidence of false positives which further increases as more patterns are added. Our method utilizes the same recovery mechanism as FRCL, but relies on a powerful, accurate and automated mechanism for bug detection.

In contrast to our checker which targets a specific class of errors, on-chip assertion processors [6] provide a more general solution, checking arbitrary properties on the chip at runtime. However, assertions must be selected on a per-design basis, thus the use of this solution requires a lot of engineering effort. Our proposed method targets a common class of design bugs, while requiring minimal effort for design time integration.

The use of hardware checkers has found an application supporting design-time formal verification. To this end, Bayazit and Malik have suggested a hybrid strategy [2] that combines hardware checkers with model checking. Their technique would allow an engineer to verify distributed properties too complex or distributed to be verified by a runtime checker, such as cache coherence in a multiprocessor system.

3. Chico Overview

Chico is an on-chip runtime checker designed to verify correct operation of the forwarding and control logic blocks in out-of-order processors. The checker is embedded in the processor design before instructions are committed. Figure 1 shows a schematic of a generic out-of-order processor design augmented with Chico. The checker adds two extra pipeline stages to the system, a Setup and a Compare stage, in addition to an extra "golden" register file where only validated register values are stored.

Chico is placed between the execute and the commit stages, thus enabling complete recovery of the processor before any architectural state is affected or erroneous results can propagate outside the core. Each instruction passing through the pipeline is checked before it can move on to commit or write its result to memory or the register file. Chico performs a series of checks on each instruction. First, it verifies that the values of each source operand match the last committed value in the golden register file. Additionally, the PC of each instruction is checked against the following instruction's PC to ensure correct program order. Should a major error cause the core to hang (deadlock), failing to commit an instruction within a specified number of clock cycles, a watchdog timer will ensure forward progress. If a failure is detected, an exception is raised and the processor switches to a *degraded mode* which is a fully functional, but barebone version of the system, obtained by disabling all units or features which boost the performance of the system, including pipelining itself. The benefit of this mode of operation is that it is typically simple enough to be formally verified, trading performance for correctness. While in degraded mode, pipelining is disabled and the failed instruction is run through by itself, thus avoiding potentially negative interactions with other instructions that would be in flight at the same time. Because the system's functionality in degraded mode has been completely verified, the instruction will compute the correct result, although extra cycles will be required for it to commit. After the failed instruction passes the checker, the normal mode of operation is re-enabled.

3.1 Strengths and Limitations

Chico was designed to overcome and correct errors occurring as a result of functional design bugs in the data forwarding and control logic portions of the processor. Because of its design, it provides the added benefit of protecting the system against transient errors occurring in these units of the core. Chico also protects against permanent transistor failures in the control units, however protection against this

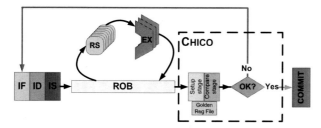

Figure 1: Chico integrated in an out-of-order pipeline. Chico checks each instruction before it proceeds to the commit stage. If the check fails, the system switches to degraded mode and re-executes the faulty instruction. The Chico hardware checker consists of two additional pipeline stages, Setup and Compare, and a golden register file to store the verified-correct register values.

latter type of error comes at a high performance cost because the recovery mode would be triggered very often. Finally, Chico can be used to shorten the time between first tapeout and customer release by facilitating the localization of control bugs on test chips between tapeout revisions.

Note, however, that the checker does not recompute the results of any instruction, it simply checks that the control information is correct. This is a consequence of assuming that core datapath components, such as adders, multipliers, logic functional units, etc., are correct. This assumption was made after observing that the verification of the processor's datapath benefits from a more mature methodology, and it is more successful in practice because it can rely on a well-defined specification and on several product generations delivering equivalent functional units. This observation is corroborated by the large majority of escaped bugs that are found in control logic blocks[9]. Because of our assumption we can maintain a very simple and lightweight design for our Chico checker, focusing exclusively on checking control logic activity. Note in addition that the Chico checker is only responsible for checking the correctness of the system's execution, but not recomputing the correct results in case of error – the degraded mode will take care of that. This aspect further simplifies the design of Chico, in fact, it has been argued on several occasions in the literature that checking a result is a simpler task than computing it[2].

The specific set of design properties we rely upon are listed below. Note that all of them are fairly straightforward and can be formally verified even in a complex design, since they involve few sequential elements. Arithmetic units such as multipliers can also be verified by checking their functional equivalence with a simple and correct implementation of the same function.

1. ALUs and other arithmetic units compute the correct result, for any given set of input operands.

2. The datapath blocks are correctly connected, that is, in absence of any dependency between instructions, execution proceeds correctly. This is equivalent to saying that the execution of individual instructions produces the correct result.

3. The memory subsystem works properly, that is, given a (data, address) pair, the memory is accessed at the proper address and the corresponding data is written or retrieved.

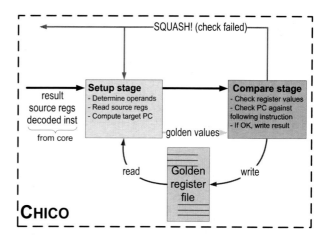

Figure 2: Components of the Chico checker. The checker is comprised of two pipeline stages, Setup and Compare, and an additional "golden" register file. The Setup stage reads the golden register file and determines which source registers need to be checked, while the Compare stage compares these values with the actual values. The Figure also shows the signals that are exchanged between the checker blocks.

Chico can also detect and correct permanent and transient faults that occur in covered blocks. Transient faults are dealt with by simply invoking the degraded mode of the system and recovering the instruction that suffered the fault. Permanent faults are different in that they cause errors which continue to manifest themselves even after re-executing the instruction in degraded mode. The risk is that the system may enter an infinite loop in which it keeps re-executing the same instruction with no forward progress. Hence, if an instruction is deemed to be erroneous even after being re-executed, Chico will raise an exception. The rationale behind this design choice was that if we let the instruction commit even when an error was detected after re-execution, the erroneous instruction would write an incorrect result with high probability, and this would undermine the integrity of the system.

3.2 Checker Design

Chico is divided into two pipelined stages, Setup and Compare, which are inserted in the pipeline directly before the commit stage. In addition, we inserted a small architected register file, which is read in the Setup stage and written in the Compare stage. Figure 2 shows a schematic of the checker blocks and the signals that they exchange. In this section, we present the architecture of the two stages and we show that this two-stage design is needed to check the correctness of the dynamic program flow.

Checker Setup Stage

The Setup stage of the checker retrieves the source register values from the checker's private architected register file (the golden register file) using the decoded instruction's register indices. It would have been possible to design the checker so that the committed register state was accessed instead by indexing through a private, golden retirement rename table into the physical register file. Since a rename table would require nearly as much space as an architected register file, we opted for the register file. This approach is also safer, as it does not rely on the correctness of the retirement rename

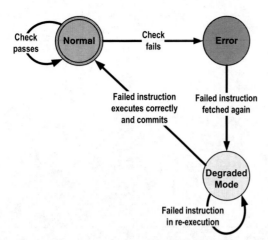

Figure 3: Pipeline control FSM. This FSM resides in the Checker Setup Stage and controls the execution flow when a control error is detected in an instruction that is about to commit. It squashes the instructions in flight and takes control of fetching instructions until the problematic instruction has been re-executed.

table. Keeping a separate register file reduces the checker complexity and guards against possible errors.

The Setup stage includes logic to determine which operands must be checked. A multiplexer produces the expected PC of the following instruction, either the branch target (when the instruction is a taken branch) or the next PC (PC+4). The Setup stage is also home to a simple finite state machine (FSM) that controls the execution flow when an error is detected, squashing all instructions in the pipeline, passing the PC of the instruction to be re-run to the fetch stage and disabling any instruction fetching until the problematic instruction has left the out-of-order core (Figure 3).

Checker Compare Stage

The second checker stage is responsible for determining if the instruction executed with the correct source register values, if the program counter corresponding to the instruction was the correct one, and finally monitors the occurrence of deadlocks. The values of the source registers are compared against those retrieved in the golden register file. The comparison is only performed on the source registers specified by the instruction being checked. The PC of the instruction currently in the Compare stage is compared against the expected value computed in the previous stage. This verifies that branches have been resolved properly and no instructions have been skipped in the dynamic flow.

Note that we can only check the correctness of the instruction flow by comparing two successive instructions' program counters. We do this by comparing the PCs of the instruction in Setup with the one in Compare. It is possible that the Setup stage could be empty, due to an earlier stall or squash. In this case, we hold on to the instruction in Compare until a new instruction arrives in Setup, then perform the PC check and release the instruction in Compare.

If both the registers and PC checks are successful, the instruction's result is written to the golden register file. Otherwise, a squash signal is asserted, flushing the entire pipeline up to (and including) the instruction in the Compare stage. This signal also causes the pipeline control FSM in the Setup stage to activate the degraded mode (see Figure 3). The PC register in the fetch stage is reset to match that of the faulty instruction and the pipeline control FSM ensures that the instruction passes through the pipeline alone. When the re-executed instruction reaches the Setup stage again, the FSM transitions out of degraded mode, enabling the fetching of the next instruction and resuming normal execution.

3.3 Integrating the Checker

Integrating Chico into a pipeline is a relatively straightforward process. With reference to Figure 1, the checker is inserted between the reorder buffer and the commit stage (or between memory and writeback stages in an in-order pipeline). The interconnects in the pipeline must be augmented to pass along source register indices, decoded register and result values, and the PC. Often, this extra interconnect is already available in the processor's pipeline to support instruction replay after non-deterministic latencies, in which case the integration phase does not have any area impact on the design. Finally, the output of the checker indicating a failed assertion must be connected to the pipeline's exception handler. Note that integration into an in-order pipeline requires consideration of data forwarding paths since additional stages are being inserted.

4. Verification Methodology

The verification of a component designed to verify another component is a challenging task. We took an incremental approach to verifying Chico, with the ultimate goal of integrating it into a complex out-of-order processor. Our testing process was divided into three phases: unit-level testing, testing in a simple in-order pipeline and final integration. In addition, the degraded mode of execution must be formally verified to be fully correct.

For each checker stage, we created a set of directed tests to verify the basic functionality at the module level. The test development was independent of the checker implementation, and based only on its specification, so as to achieve maximum separation between implementation and testing. Most of the errors encountered during this early phase of testing were related to instruction semantics, particularly in determining which registers were actually accessed by each instruction, and hence should be checked by the checker.

In order to test our checker before integrating it into the out-of-order core, we inserted it in a simple 5-stage in-order pipeline. This was valuable in that it helped us work out any implementation-related bugs that might not have been caught with isolated testing. We randomly inverted control signals while running a large set of test programs, comparing the values written back on commit to those produced by a known-good processor model. An unexpected challenge arose here when we realized that adding two stages before memory and writeback would require extra forwarding logic. Conveniently, the checker helped in its own integration by pointing out errors in the new forwarding logic.

Final integration of Chico into the out-of-order processor design was relatively uneventful as a result of the thorough testing carried out beforehand. The pipeline's exception handling logic had to be slightly altered to handle Chico's squash signal, and the system's fetch enable was connected to the checker's pipeline control FSM.

The degraded mode of operation requires little alteration to the processor core into which Chico is being integrated. It is important to note that degraded mode does not require a new simplified version of the design. Many non-essential units, such as branch predictors and speculative execution units can be disabled with a variant of chicken bits, which are common in many design developments. Additional simplification is achieved by virtue of the single instruction execution, only one instruction is allowed in the pipeline at a time when in degraded mode. The whole ISA is verified, one instruction at a time, allowing formal tools to abstract away forwarding, squashing and out-of-order execution logic as well as greatly reducing the fraction of the design involved in each individual property proof. The combination of disabled logic blocks and unused logic due to single instruction execution make the degraded mode simple enough to be verified by traditional formal tools.

5. Experimental Evaluation

We implemented and integrated the Chico checker in a out-of-order pipeline design, and evaluated its error-detection qualities by creating eleven variants of the processor, each with a different design error. All the errors were inspired by the bugs reported in errata documents of current processors in the market today. We measured the performance impact associated with using Chico by running three different types of testbenches on these design variants: pseudo-random testing, targeted direct test cases and the SPECint benchmark suite. These three types of tests enabled us to evaluate the performance of Chico in scenarios ranging from real-world applications to targeted extreme execution flows. In addition, we synthesized our design to estimate Chico's area impact.

A number of architectural parameters may impact the performance penalty of a chip utilizing our checker. Since the recovery mechanism invokes the built-in exception handler, squashing all instructions in flight in an out-of-order processor, the penalty due to a recovered error may vary. Pipeline depth is the most significant factor in determining this penalty. Dependencies between instructions, execution latencies (especially for loads and stores) and branch prediction accuracy all have an effect on the performance impact. We quantified the penalty for varying error rates using an RTL simulation of a moderately sized out-of-order processor. The out-of-order pipeline design used in our tests implements early branch resolution, meaning that branches are resolved in the core rather than on commit. Because of this, the penalty for mispredicted branches is not affected by the addition of Chico's two checker stages.

Issue / retire capacity	single issue/retire
Physical Registers	64
ROB entries	32
Reservation stations	16
Memory access	4 load units
Memory latency	8 cycles
Caches	128 line I-cache and D-cache
Branch prediction	64 entry hashing, 2-bit counter

Table 1: Characteristics of the out-of-order testbench processor in which we integrated Chico.

5.1 Experimental Setup

Chico was developed in Verilog HDL and integrated into a 64-bit out-of-order pipeline which implements a subset of the Alpha ISA. The subset includes about 95% of the integer operations and all of the branching instructions. Floating point, byte manipulation and multimedia instructions were excluded. We made use of three different experimental setups to execute programs in the processor, a standard fetch for running targeted assembly test cases, a connection to a constrained random stimulus generator, and a parallel lockstep connection with an architectural simulator. The architectural simulator allowed us to simulate SPECint benchmark by emulating the instructions not supported by our ISA. The features of the out-of-order processor are shown in Table 1. Note that while here we evaluate Chico on a single issue/retire design, the checker could easily be extended to multiple issue/retire by adding a dependency check among instructions retired simultaneously.

In order to evaluate the bug-detection capabilities of our checker, we made use of a set of buggy processor designs implemented in Verilog. The bugs were created to mimic escaped bugs found in errata documents for modern processors, including x86, PowerPC and ARM[3, 4, 5]. The Verilog model of our out-of-order processor was manually modified, inserting specific bugs as necessary. These bugs are described in Table 2.

Bug	Description
baseline	no bugs
two-stores	two consecutive stores cause incorrect address computation
two-branches	two consecutive branches corrupt program order
store-dep	store followed by a dependent instruction fails
mult-branch	multiply followed by a branch causes branch to resolve incorrectly
load-branch	conditional branch depending on a preceding load fails
rob-full	full reorder buffer causes disruption in program flow
rs-full	all reservation stations full causes a missed instruction
dep-instrs	two consecutive dependent instructions fail
zero-reg-CDB	forwarding from zero reg fails with simultaneous CDB broadcast
write-zero-reg	write to zero reg causes non zero values to be read
regA-regB-fwd	simultaneous regA and regB forwarding causes incorrect reg value

Table 2: Bugs introduced into the out-of-order core

5.2 Simulated Workloads

We have evaluated the effectiveness and performance of Chico with three types of workloads: constrained random stimulus, directed assembly language test cases and SPECint benchmarks.

In the first setup, instruction streams generated by a closed loop constrained random test generator[8] were feeding the instruction bus of the processor. For this simulation we used our baseline out-of-order processor design without inserted bugs. The generator was configured to exercise the full ISA of our system, stressing in particular dependencies and data forwarding through memory. This gave us additional reassurance that the checker operated as designed.

We also utilized a set of directed tests in assembly language that were developed as part of the verification process during the initial development of the processor design. These test cases are described in Table 3. Each of the testbenches runs for a moderately high number of cycles and stress all the key features of the processor, particularly the control aspects. We found experimentally that this regression suite was capable to expose all the 11 bugs that were included in the design variants. These programs, while more suited to exposing bugs, might give pessimistic estimates of performance compared to real workloads. Because the test cases were designed to maximize the number of in-flight instructions, the penalty incurred on a checker recovery also tends to be maximized.

Name	Description
bubblesort	bubble sort an array of numbers
combRec	recursively calculates combinations
dude	sorts, divides and square roots arrays
fib_rec	computes Fibonacci sequences recursively
objsort	sort items in a linked list
parsort	comparison sort an array of numbers
prime	finds all prime numbers less than X
powers	computes large powers
series	computes geometric and arithmetic series

Table 3: Targeted test cases written in assembly language

To obtain results reflecting more accurately real-world execution flows, a set of 12 SPECint benchmarks were used in our third experimental setup. These provided us with an accurate estimate of the effect of the checker recovery on CPU performance. Because the benchmarks are prohibitively large, they could not be executed in their entirety using our RTL model. To derive a shorter trace that would be representative of each SPECint benchmark, we used SimPoint. The traces generated by SimPoint are derived from the actual benchmarks, but each contain less than 10k instructions. In effect, the traces fast-forward through the initialization of each benchmark, including only the core part of the program most indicative of overall performance. This method allowed us to extract realistic performance data while maintaining a practical simulation time. The benchmarks were compiled to target the full Alpha ISA, including a few instructions not handled by our processor. To solve this problem, we ran our model in lockstep with the SimpleScalar[1] simulator. SimpleScalar would simulate the SPECint benchmark and output a dynamic trace file which we fed to our RTL processor design. The trace file was consulted by the processor when it encountered instructions that were not implemented in our design. By doing so, we were able to follow the correct execution path and also receive the proper data from emulated system calls. This experiment provided us with a more realistic evaluation of the frequency of occurrence of escaped bugs during a typical workload execution and, consequently, of the performance penalty incurred by the use of Chico.

5.3 Results

When no buggy configuration arises, our method has a negligible impact on processor performance. The additional pipeline stages inserted for the Chico checker increase instruction latency, but this has no significant effect on the instructions-per-cycle (IPC) for large test programs. We purposely designed the processor to achieve this by freeing physical registers immediately after instructions leave the ROB, and by resolving branches during execution.

Table 4 reports our experimental results over the SPECint benchmarks. The first column indicates the testbench name,

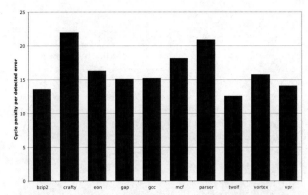

Figure 4: Average cycle penalty for each error recovery over all 11 buggy versions of the out-of-order pipeline equipped with Chico, while executing SPECint benchmarks.

the second the number of dynamic instructions executed, the third is the relative IPC, that is, the ratio between the instructions-per-cycle when executing the testbench on a buggy design version and the IPC when executing on a correct design version which has not been equipped with Chico. Finally, the last two columns report the total number of recoveries triggered by Chico and the average number of penalty clock cycles incurred during each recovery. Note that the table reports results averaged over all the 11 buggy variants of the design. Figure 4 shows graphically the average number of clock cycles of penalty incurred with each Chico recovery, the same data reported in the last column of Table 4.

When errors are present and caught by the checker, we still observe a fairly small impact on the overall IPC, particularly when running SPECint benchmarks. The slowdown in this case ranged from less than 1% to 4%.

The SPECint benchmarks provide the best workload for determining the checker's performance. In general, these programs exhibited a lower penalty per fault and degradation in the overall IPC (1% - 4%). However, the SPECint suite exposed significantly fewer bugs than the targeted test cases. Three programs, *gzip*, *mcf* and *perlbmk* were unable to expose any of the embedded bugs.

Figure 5 reports the same information as Figure 4 for the directed testbenches suite. It can be noticed that the penalty incurred is much higher in this suite compared to SPECint. The reason lies in the high density of in-flight instructions, which was achieved by developing this suite directly in as-

Benchmark	#Instr	rel IPC	Recoveries	Penalty
bzip2	8613	0.967	77	13.6
crafty	7061	0.996	33	22.0
eon	6634	0.987	60	16.3
gap	7071	0.976	118	15.1
gcc	6534	0.975	37	15.2
gzip	5985	1	0	N/A
mcf	4771	1	0	N/A
parser	6148	0.987	76	20.9
perlbmk	5137	1	0	N/A
twolf	7821	0.992	48	12.6
vortex	5551	0.995	25	15.8
vpr	6927	0.993	39	14.1

Table 4: Performance summary for SPECint benchmarks, averaged over all buggy designs, relative to baseline design.

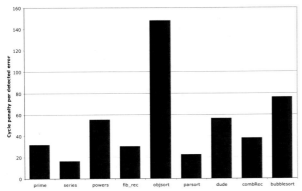

Figure 5: Average cycle penalty for each error recovery over all 11 buggy versions of the out-of-order pipeline design equipped with Chico, while executing each of the targeted test cases.

sembly. We also note that this suite was more effective at exposing bugs than the SPEC benchmarks. The *objsort* program is an example of nearly worst-case behavior, with an exceptionally high penalty for checker recovery. In this program, a large number of back-to-back dependent memory operations may be flushed from the pipeline whenever an error is encountered. The overall slowdown with directed test cases is greater than with the SPECint suite, with up to 30% performance impact in the worst case.

Simulation with the random stimulus generator reinforced our confidence in Chico's ability to differentiate bugs from correct operation. We ran nearly 12 million dynamic instructions using this setup, finding that Chico flagged no false positives. Interestingly, Chico exposed a few escaped bugs in our processor core which had not been caught during verification when it was originally designed by a team of graduate students. The results from our direct random stimulus tests reassured us that Chico was free from false positives and operated as designed.

5.4 Area Cost

We estimated the area overhead of our Chico implementation by synthesizing the experimental processor design with and without the inclusion of Chico. To this end, we used Synopsys Design Compiler targeting a 90nm TSMC library. Design Compiler was configured with timing as the primary objective and area as the second; a basic wire delay model from the library was used. The resulting checker area was $0.065mm^2$, while the processor core occupied $2.291mm^2$; thus Chico comprises 2.8% of the total area. Note that 2.8% is a very conservative area penalty due to the small size of the processor used for experimentation; a larger design, for example one that includes a floating point unit, a load-store queue, or other features, would further reduce Chico's relative size. Chico's total area can be broken down into its three components, with 5% comprised of the Setup stage, 2% of the Compare stage and 93% of the architected register file.

For comparison, we used the same methodology to estimate the area of a DIVA[10] solution in 90nm technology. Our DIVA implementation was a simple, single-issue 5 stage pipeline with a 0.5KB instruction cache and a 4KB data cache very similar to the one described in [10]. We synthesized the DIVA processor core with the same 90nm TSMC library as was used for Chico and approximated the cache areas with Cacti 4.2[7]. Note that this is a conservative estimate, as it is only the area of the DIVA checker processor itself and does not include any of the extremely complex interconnects necessary to connect each stage of the checker processor to its corresponding stage in the processor core. Our estimated area for a single-issue DIVA processor is $0.673mm^2$, about 10 times the area of Chico.

6. Conclusions

We have presented Chico, an efficient approach to on-chip runtime verification that specifically targets functional errors in the control logic blocks of a processor. The area penalty of our Chico implementation is 10 times less than an approach based on a full checker processor. By taking advantage of a formally verified degraded mode, our solution recovers from functional errors in exchange for a small performance impact. Our solution is novel in that it targets a critical class of design bugs and optimizes its design based on the characteristics of this class of bugs. The error detection logic does not trigger false positives, resulting in zero performance degradation when no bugs are encountered. We show an average performance penalty ranging from less than 1% to 4% for the SPECint benchmarks running on a range of buggy processor designs. Our checker is effective in converting correctness concerns into performance issues, and it supports the verification team in managing the complexity of large out-of-order CPUs, increasing reliability and easing the verification process.

7. References

[1] T. Austin, E. Larson, and D. Ernst. Simplescalar: An infrastructure for computer system modeling. *IEEE Computer*, 35(2):59–67, February 2002.

[2] A. Bayazit and S. Malik. Complementary use of runtime validation and model checking. In *International Conference on Computer-Aided Design*, pages 1052–1059, November 2005.

[3] DDJ Microprocessor Center. http://www.x86.org/.

[4] IBM Corporation. IBM PowerPC 750GX and 750GL RISC Microprocessor Errata Notice, July 2005.

[5] Intel Corporation. Intel(R) StrongARM(R) SA-1100 Microprocessor Specification Update, February 2000.

[6] J. Nacif, F. de Paula, H. Foster, C. Coelho, and A. Fernandes. The Chip is Ready. Am I done? On-chip Verification using Assertion Processors. In *Symposium on Integrated Circuits and System Design*, pages 55–59, September 2004.

[7] D. Tarjan, S. Thoziyoor, and N. Jouppi. Cacti 4.0, June 2006.

[8] I. Wagner, V. Bertacco, and T. Austin. Stresstest: An automatic approach to test generation via activity monitors. In *Design Automation Conference*, June 2005.

[9] I. Wagner, V. Bertacco, and T. Austin. Shielding against design flaws with field repairable control logic. In *Design Automation Conference*, pages 344–347, July 2006.

[10] C. Weaver and T. Austin. A fault tolerant approach to microprocessor design. In *International Conference on Dependable Systems and Networks*, pages 411–420, 2001.

A CLP-based Functional ATPG for Extended FSMs*

Franco Fummi† Cristina Marconcini† Graziano Pravadelli†
†Department of Computer Science
University of Verona, Italy
Strada Le Grazie 15, 37134 Verona, Italy
{franco.fummi,cristina.marconcini,graziano.pravadelli}@univr.it

Ian G. Harris‡
‡Department of Computer Science
University of California Irvine
Irvine, CA, USA
harris@ics.uci.edu

ABSTRACT

It is a common opinion that semi-formal verification offers a good compromise between speed and exhaustiveness. In this context, the paper presents a semi-formal functional ATPG that joins static and dynamic techniques to generate high-quality test sequences. The ATPG works on a set of concurrent extended finite state machines (EFSMs) that models the design under verification (DUV). The test generation procedure relies on backjumping, for traversing the EFSM transitions, and constraint logic programming (CLP), for covering corner cases through the deterministic propagation of functional faults observed, but not detected, during the transition traversal.

Keywords

Functional verification, ATPG, EFSM, CLP.

1. INTRODUCTION

Functional ATPGs based on simulation [1, 2] are fast, but generally, they are unable to cover corner cases, and they cannot prove untestability. On the contrary, functional ATPGs exploiting formal methods [3, 4, 5], being exhaustive, cover corner cases, but they tend to suffer of the state explosion problem when adopted for verifying large designs. In this context, the paper proposes a functional ATPG that relies on the joint use of pseudo-deterministic simulation and constraint logic programming [6], following the promising trend of adopting semi-formal approaches for solving complex problems. Thus, the advantages of both simulation-based and static-based verification techniques are preserved, while their respective drawbacks are limited. In particular, CLP, a form of constraint programming in which logic programming is extended to include concepts from constraint satisfaction, is well-suited to be jointly used with simulation. In fact, information learned during design exploration by simulation can be effectively exploited for guiding the search of a CLP solver towards DUV areas not covered yet. Therefore, this paper is focused on the use of CLP for addressing corner cases during functional test pattern generation. In particular, a CLP-based fault-oriented ATPG engine is proposed to be adopted, after simulation, learning and backjumping, as the last step of the incremental test generation flow showed in Figure 1.

According to such a flow, the ATPG framework is composed of three functional ATPG engines working on three different models of the same DUV: the hardware description language (HDL) model of the DUV, a set of concurrent EFSMs extracted from the HDL description, and a set of logic constraints modeling the EFSMs. The EFSM paradigm has

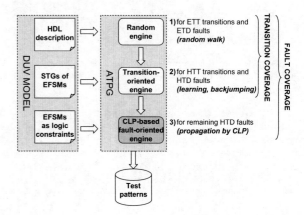

Figure 1: The incremental test generation flow.

been selected since it allows a compact representation of the DUV state space [7], that limits the state explosion problem typical of more traditional FSMs.

The first engine is random-based, the second is transition-oriented, while the last, which represents the main contribution of the paper, is fault-oriented. The test generation is guided by means of transition coverage [8] and fault coverage [9]. In particular, 100% transition coverage is desired as a necessary condition for fault detection, while the bit coverage [10] functional fault model is used to evaluate the effectiveness of the generated test patterns by measuring the related fault coverage.

A random engine is first used to explore the DUV state space by performing a simulation-based random walk. This allows us to quickly fire easy-to-traverse (ETT) transitions and, consequently, to quickly cover easy-to-detect (ETD) faults. However, the majority of hard-to-traverse (HTT) transitions remain, generally, uncovered. Thus, a transition-oriented engine is applied to cover the remaining HTT transitions by exploiting the learning/backjumping-based strategy described in [11]. Backjumping, also known as non-chronological backtracking, is a special kind of backtracking strategy which rollbacks from an unsuccessful situation directly to the cause of the failure [12]. Thus, the transition-oriented engine works as follows:

1. it deterministically backjumps to the source of failure when a transition, whose guard depends on previously set registers, cannot be traversed;

2. it modifies the EFSM configuration to satisfy the condition on registers;

3. it successfully comes back to the target state to activate the transition.

---
*This work has been partially supported by the VERTIGO European project FP6-2005-IST-5-033709.

The transition-oriented engine generally allows us to achieve 100% transition coverage. However, 100% transition coverage does not guarantee to explore all DUV corner cases, thus some hard-to-detect (HTD) faults can escape detection preventing the achievement of 100% fault coverage. Therefore, the CLP-based fault-oriented engine is finally applied to focus on the remaining HTD faults.

The rest of the paper focuses on such a CLP-based fault-oriented engine. In particular, it shows how:

- a set of EFSMs can be modeled by CLP; this is not a trivial task, since modeling the evolution in time of an ESFM by using logic constraints is really different with respect to model the same behavior by means of a traditional HW description language;
- a CLP solver can be used to deterministically search for sequences that propagate the HTD faults observed, but not detected, by the random and the transition-oriented engines;
- the risk of state explosion, arising from the use of CLP, can be limited by some strategies that allow us to prune the search space.

The paper is organized as follows. Section 2 is devoted to related works. Section 3 summarizes the main concept related to EFSMs and it presents how they can be modeled by CLP. Section 4 describes the use of CLP within the fault-oriented engine. Finally, Section 5 reports experimental results, and Section 6 summarizes the concluding remarks.

2. RELATED WORKS

CLP-based ATPGs have been already proposed in the literature [13, 14, 15, 16, 11], however, existent approaches differ in several aspects from the ones presented in this paper.

In [13], CLP is used to generate test sequences according to a path coverage-based criterion. However, this approach, is oriented only to the control part of circuits. Control flow paths are also the target of the approach presented in [14, 15]. In [14], constraints are derived from a preprocessing of a VHDL description to enumerate all the target paths. On the contrary, in [15], constraints are generated by enumerating paths of concurrent FSMs describing the DUV. However, path enumeration is a very hard and time-consuming task, since paths of sequential circuits are generally infinite. In [16], the authors propose to use CLP for generating test sequences targeting synchronization/timing faults in HW/SW models described as a network of co-design FSMs. This work identifies sequences to trigger synchronization faults but the observability of the fault effect is not considered.

A CLP-based ATPG for traversing EFSMs has been proposed in [11]. However, in such a work CLP is exploited only to deterministically activate the guard of HTT transitions, after that the learning/backjumping mechanism has dynamically generated the opportune constraints to be solved. Moreover, test generation is guided by transition coverage, thus, preventing the identification of corner cases that can be accomplished by using a fault model.

Finally, the previous approaches propose neither strategies for completely modeling a design by means of CLP (just some paths are modeled), nor approaches for avoiding the risk of state explosion when a CLP solver is asked to generate a test sequence to activate the target path.

3. COMPUTATIONAL MODEL

In this paper, we represent a digital system as a set of concurrent EFSMs, one for each process of the DUV. In this way, according to the below Def. 1, we capture the main characteristics of state-oriented, activity-oriented and structure-oriented models [17]. In fact, the EFSM is composed of states and transitions, thus it is state-oriented, but each transition is extended with HDL instructions that act on the DUV registers. In this sense, each transition represents a set of activities on data, thus, the EFSM is a data-oriented model too. Finally, concurrency is intended as the possibility that each EFSM of the same DUV changes its state concurrently to the other EFSMs to reflect the concurrent execution of the corresponding processes. Data communication between concurrent EFSMs is guaranteed by the presence of common signals. In this way, structured models can be represented.

DEFINITION 1. *An EFSM is defined as a 5-tuple $M = \langle S, I, O, D, T \rangle$ where: S is a set of states, I is a set of input symbols, O is a set of output symbols, D is a n-dimensional linear space $D_1 \times \ldots \times D_n$, T is a transition relation such that $T : S \times D \times I \rightarrow S \times D \times O$. A generic point in D is described by a n-tuple $x = (x_1, ..., x_n)$; it models the values of the registers of the DUV. A pair $\langle s, x \rangle \in S \times D$ is called configuration of M.*

An operation on M is defined in this way: if M is in a configuration $\langle s, x \rangle$ and it receives an input $i \in I$, it moves to the configuration $\langle t, y \rangle$ iff $((s, x, i), (t, y, o)) \in T$ for $o \in O$.

The EFSM differs from the classical FSM, since each transition does not present only a label in the classical form $(i)/(o)$, but it takes care of the register values too. Transitions are labeled with an *enabling* function e and an *update* function u defined as follows.

DEFINITION 2. *Given an EFSM $M = \langle S, I, O, D, T \rangle$, $s \in S, t \in T, i \in I, o \in O$ and the sets $X = \{x | ((s, x, i), (t, y, o)) \in T \text{ for } y \in D\}$ and $Y = \{y | ((s, x, i), (t, y, o)) \in T \text{ for } x \in X\}$, the enabling and update functions are defined respectively as:*

$$e(x, i) = \begin{cases} 1 & \text{if } x \in X; \\ 0 & \text{otherwise.} \end{cases}$$

$$u(x, i) = \begin{cases} (y, o) & \text{if } e(x, i) = 1 \text{ and} \\ & ((s, x, i), (t, y, o)) \in T; \\ undef. & \text{otherwise.} \end{cases}$$

An update function $u(x, i)$ can be applied to a configuration $\langle s_1, x \rangle$ if there is a transaction $t : s_1 \rightarrow s_2$, labeled e/u, such that $e(x, i) = 1$. In this case we say that t can be *fired* by applying the input i. Figure 2 shows the STG of a simple EFSM.

3.1 EFSM Generation

Many EFSMs can be generated starting from the same HDL description of a DUV. However, despite from their functional equivalence, they can be more or less easy to be traversed. Easiness of traversal is a mandatory feature for a computational model used in test pattern generation, and it is a desirable condition to activate and propagate faults.

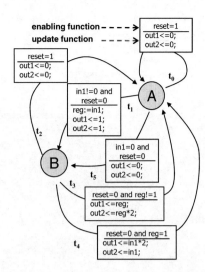

Figure 2: STG of a simple EFSM.

Stabilization of EFSMs improves the easiness of traversal [7], but it can lead to the state space explosion. Thus, in [18], a set of theoretically-based automatic transformations has been proposed to generate a particular kind of semi-stabilized EFSM (S^2EFSM). It allows an ATPG to easily explore the state space of the corresponding DUV reducing the risk of state explosion. The S^2EFSM presents the following characteristics:

- It is functionally and timing equivalent to the HDL description from which it is extracted, i.e., given an input sequence, the HDL description and the corresponding S^2EFSM provide the same result at the same time.

- The update functions contain only assignment statements. This implies that all the control information, needed by a deterministic ATPG to traverse the DUV state space, resides in the enabling functions of the S^2EFSM.

- The S^2EFSM is partially stabilized to reduce the state explosion problem that may arise when stabilization is performed to remove inconsistent transitions. Only transitions not leading to state explosion are stabilized.

3.2 EFSM Modeling by CLP

The fault-oriented engine needs a CLP-based representation of the DUV, and some searching functions to generate test sequences as reported in Section 4. The CLP-based representation is automatically derived from the S^2EFSM models according to the rules reported in the next subsections, which follow the syntax of the ECLiPSe CLP solver [19].

At first, we introduce the concept of time steps, required to model the S^2EFSM evolution through the time via CLP. Then, we deal with modeling of logical variables and constraints to represent enabling functions and update functions of the S^2EFSM. The search functions required for generating test sequences are presented in Section 4.2.

3.2.1 Modeling of Time

Hardware description languages easily allow to model the DUV evolution in time by means of processes, implicit or explicit wait statements, sensitivity lists and events. On

```
efsm(Max)  :-
N #< 50, N #> 1,
N :: 1..Max, indomain(N),
% Definition of logical variables as arrays of size N
% Definition of the DUV behavior by using logical variables
% Searching functions
```

Figure 3: Skeleton of the CLP description for time modeling.

the contrary, CLP does not provide an explicit mechanism to model the time evolution. To overcome this limitation, we introduce a logical variable, N, to represents the total number of time frames on which the S^2EFSM can evolve. The range of N is [1,Max], where Max is defined according to the sequential depth of the DUV. Then, the *CLP variables*, used to model the S^2EFSM behavior, are defined as *arrays* of size N. Thus, for example, let us consider a variable V defined in the HDL description of the DUV. When CLP is adopted, an array V[] is used to model the evolution in time of variable V. In this context, a CLP constraint of the form V[T] #= 0 indicates that the element of index T of the array V[] (in the following referenced as Tth element) equals 0. Such a constraint is used to specify that at time T the variable V of the DUV has value 0.

A further constraint is added to ask the CLP solver to stop searching, if a solution exists, before N reaches the Max value. Otherwise, all sequences generated by CLP would be of length Max, thus, requiring the solver to uselessly work even if shorter sequences exist. Figure 3 shows how the solver is invoked for searching the shortest sequence whose length is included in the range [1,Max].

3.2.2 Modeling of Variables

Three kinds of arrays of logical variables must be defined to describe, respectively, states, transitions, and signals/registers of an S^2EFSM.

State variables. Each state of the S^2EFSM is modeled by an array of boolean variables of size N. When the EFSM is in the state S at time T, the Tth element of the array S[], corresponding to the state S, is assigned to true. For example, two arrays, A[] and B[], are required to model states A and B of the S^2EFSM of Figure 2. At every time step T, either A[T] or B[T] is true, thus indicating the current state of the S^2EFSM.

Transition variables. Each transition of the S^2EFSM is associated to an array of boolean variables of size N. If a transition is fired at time T, the Tth element of the array corresponding to the transition is assigned to true.

Signal/register variables. Signals and registers are modeled as arrays of size N respecting their original data type.

The CLP code of Figure 4 exemplifies the constraints required to model logical variables for states, transitions, signals and registers of the S^2EFSM of Figure 2.

The predicate bool(X,N) defines the boolean data type used to model states and transitions. It means that the logical variable X is an array of size N (for modeling of time), whose elements can assume the values included in the list Xlist, i.e., 0 or 1. In a similar way, the predicate int32(X,N) defines an arrays of N 32-bit integers used as data type to deal with PIs, POs and registers.

```
% data types
bool(X,N):- dim(X,[N]),term_variables(X, Xlist),
            Xlist::[0,1].
int32(X,N):- dim(X,[N]),X[1..N]::-2147483648..2147483647.
% states
bool(A,N),bool(B,N),
% transitions
bool(T0,N),bool(T1,N),bool(T2,N),bool(T3,N),
bool(T4,N),bool(T5,N),
% PIs, POs and registers
int32(IN1,N),int32(REG,N),int32(OUT1,N),int32(OUT2,N)
```

Figure 4: Constraints for modeling states, transitions, signals and registers of the S^2EFSM of Figure 2.

3.2.3 Modeling of States and Transitions

The functional behavior of the S^2EFSM is represented by means of enabling functions and update functions labelling the transitions between states. Thus, a way for modeling such functions and their relation with states and transitions is proposed. In particular, two kinds of constraints have been defined to model the current state of the S^2EFSM, and the relation between the enabling function and the corresponding update function.

Current state. Two constraints must be defined for each array of state variables to specify the current state of the S^2EFSM. The first constraint specifies that, at each time step T, the Tth element of an array S[], modeling a state S of the DUV, is true, if and only the Tth element of one of the transition arrays corresponding to the transitions in-going in S is true. The second constraint specifies the dual situation, i.e., if the Tth element of the transition array is true at time T, then the Tth element of the array associated to the destination state of the corresponding transition must be true at time step $T + 1$ (NEXT_T). For example, let us consider the S^2EFSM of Figure 2. The following constraints must be defined for specifying that the current state of the S^2EFSM at time step $T + 1$ is A, if and only if one of the transitions in-going in A has been fired at time T.

```
A[NEXT_T] #= T0[T] xor T1[T] xor T2[T] xor T3[T]
            xor T4[T] xor T5[T],
T0[T] xor T1[T] xor T2[T] xor
T3[T] xor T4[T] xor T5[T] => A[NEXT_T].
```

Finally, a further constraint is introduced to explicitly force the system to be in a single state at each time step. Thus, the Tth element of arrays corresponding to states of the S^2EFSM are put in *xor* each other as follows:

```
A[T] xor B[T].
```

The same constraint is define also for the variable set corresponding to the EFSM's transitions. For example:

```
T0[T] xor T1[T] xor T2[T] xor T3[T] xor T4[T] xor T5[T].
```

In fact, at a particular time step, only one transition of the S^2EFSM can be traversed and, obviously, the S^2EFSM can have only one state active. Designers implicitly include such a constraint, when they model the DUV by means of an HDL. However, the explicit presence of such a constraint, when the S^2EFSM is provided to the CLP solver, allows the solver to immediately prune the solution space by ignoring configurations where more than one state variable is concurrently true, thus drastically reducing the number of backtracking steps.

Enabling and update functions. Firing a transition at time T implies that: (i) its enabling function is satisfied at time T, (ii) its update function is executed at time T, and (iii) the state of the S^2EFSM at time T is the source of the transition. Thus, for example, if Ti is a transition out-going from state S, whose enabling function and update function are modeled, respectively, by the predicates EF and UF (described later), the following constraints are used to model Ti.

```
EF[T] and S[T] => Ti[T],
Ti[T] => EF[T] and S[T],
Ti[T] and (EF[T] and S[T]) => UF[T]
```

The first two constraints represent a double implication for imposing that the transition variable Ti[T] is true (i.e., the transition Ti is fired at time T) if and only if the predicate of the corresponding enabling function EF[T] and the variable S[T], associated to the state from which Ti is out-going, are true. On the contrary, the predicate of the update function UF[T] does not require a double implication, because it is possible that UF[T] is true even if EF[T] is false. However, in this case the transition is not fired and the update function is not executed.

The predicate EF[T] is directly derived from the condition involved in the corresponding enabling function. Its modeling requires only a syntactical conversion from the syntax of the HDL used to model the DUV towards the syntax accepted by the CLP solver.

On the contrary, modeling the predicate UF[T] associated to an update function requires more attention. In particular, an update function involves assignments to signals, registers and POs. Let us use an example to show how to model such a kind of statement. Consider, for example, the statement SIG := SIG + IN, where SIG is an internal signal and IN is a primary input. The corresponding CLP constraint is SIG[Next_T] #= SIG[T] + IN[T].

However, signals, registers and POs, that do not require to be updated, are not assigned in the update function when a design is modeled by using an HDL. Indeed, they implicitly preserve their previous value. Unfortunately, the CLP solver assigns random values to variables that are not explicitly assigned. Thus, when an update function is modeled by means of constraints, we have to ensure that a constraint is explicitly added to preserve the value of signals, registers and POs that do not require to be updated.

According to the previous rules, for example, the update function of transition t_4 in Figure 2 is modeled as follows:

```
((REG[T] #= 1) and B[T]) => T4[T],
T4[T] => ((REG[T] #= 1) and B[T]),
    (T4[T] and ((REG[T] #= 1) and B[T]) =>
    ((REG[Next_T] #= REG[T]) and
    (OUT1[Next_T] #= IN1[T]*2) and
    (OUT2[Next_T] #= IN1[T])).
```

4. CLP-BASED ENGINE

The transition-oriented engine described in [11] pseudo-deterministically generates sequences for firing HTT transitions on S^2EFSMs. In this way, the majority of faults are detected as a consequence of transition traversal, but some HTD faults can remain uncovered. On the contrary, the CLP-based fault-oriented engine exhaustively searches for test sequences targeting specific faults. It exploits the CLP-solver to explore the CLP-based representation of the DUV extracted from the S^2EFSM model. The exhaustiveness, guaranteed by CLP, is paid in terms of execution time, but such an engine is applied to a small number of faults: those not detected neither by the random-based engine nor by the transition-oriented one.

Let us consider a fault f that has not been detected yet by these engines. This may depends on two different reasons:

1. the ATPG has been unable to find an *activation sequence*, i.e., in the case of the bit coverage fault model, a sequence that causes the bit (or the condition) affected by f to be set with the opposite value with respect to the one induced by f;

2. the ATPG activated f, but it has been unable to find a *propagation sequence*, i.e., a sequence that propagates the effect of f to the primary outputs (POs) of the DUV.

To distinguish between the previous alternatives, the simulator integrated in the ATPG observes the effect of each fault on both POs and internal registers, during the simulation of test sequences generated by the random-based and transition-based engines. From this observation, the following well-known definitions derive.

DEFINITION 3. *A fault* f *is said to be observable on POs, i.e., detectable, if there exists a test sequence s such that, by concurrently applying s to the faulty and the fault-free DUVs, the value of at least one PO in the fault-free DUV differs from the value of the corresponding PO in the faulty DUV, at least once in time.*

DEFINITION 4. *A fault* f *is said to be observable on a register if there exists a test sequence s such that, by concurrently applying s to the faulty and the fault-free DUVs, the value of at least one register in the fault-free DUV differs from the value of the corresponding register in the faulty DUV, at least once in time.*

According to the previous definitions, if the fault is observed on POs, it is marked as detected and the corresponding test sequence is saved. Otherwise, if the fault is observed only on registers, the fault is marked as *to be propagated* (TBP). Finally, if the fault is observable neither on POs nor on registers, this is due to the difficulty of finding an activation sequence. Thus, the fault is marked as *to be activated* (TBA).

The current version of the CLP-based fault-oriented engine addresses TBP faults. The next subsection describes how such an engine allows us to find propagation sequences for TBP faults (see Figure 5). Future works will address the problem of TBA faults.

4.1 Propagation Sequence

The propagation sequence for a TBP fault is generated by providing the ATPG engine with two instances of the CLP-based representation of the DUV. These are initialized with the S^2EFSM configurations (the faulty and the fault-free ones) that allow the random or the transition-based

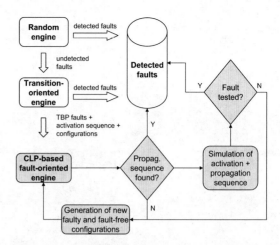

Figure 5: The role of the CLP-based fault-oriented engine.

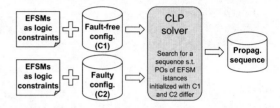

Figure 6: Use of the CLP solver for finding propagation sequences.

engine to observe the fault on at least one register[1]. Such configurations consist of the values of registers (included the value of the state register) in the faulty and fault-free DUVs at the moment the fault has been observed. As an example, consider the following constraints:

```
% ------------ FAULT FREE ------------
REG[1] #= 5941, A[1] #= 1, B[1] #= 0,
% ------------ FAULTY ------------
REG_F[1] #= 5941, A_F[1] #= 0, B_F[1] #= 1.
```

They state that, at time $T = 1$, REG has the same value (i.e., 5941) in both the faulty and fault-free DUV instances, but the two S^2EFSMs are in different states (i.e., the fault-free DUV is in state A, while the faulty DUV is in state B).

Moreover, at time $T = 1$, we force the POs of the faulty and fault-free DUVs to be equal. This constraint is mandatory to avoid the CLP solver considers the POs as free variables to be randomly initialized, which definitely would lead to a wrong solution. Thus, constraints like the following, for the example of Figure 2, must be specified:

```
OUT1[1] #= OUT1_F[1], OUT2[1] #= OUT2_F[1].
```

---

[1] Please, note that the two DUV instances provided to the CLP solver are exactly equal, but their registers are initialized with different values (the faulty and the fault-free ones). This is the only reason such instances should behave in different ways. In the following of the paper, we use the terms *faulty* and *fault-free* to distinguish the DUV instances initialized, respectively, with the faulty and the fault-free configuration.

After the set-up of the faulty and fault-free configurations, the CLP-solver is asked to find a sequence that, starting from such configurations, propagates the effect of the fault towards the POs (see Figure 6). Therefore, we define a constraint that force the arrays of POs to be different at least in a position as follows:

`~((OUT1 = OUT1_F) and (OUT2 = OUT2_F))`

where $\sim$ in ECLiPSe is the sound negation operator. If the solver finds a solution, it consists of a propagation sequence that can be appended to the activation sequence, previously generated by the random-based or transition-based engine to observe the target fault on the internal registers. The so obtained sequence is very likely a test sequence for the target fault, but this must be definitely proved via simulation (see Figure 5). In fact, the propagation sequence is generated by initializing a fault-free instance of the DUV with a faulty configuration, which is not the same as using a real faulty DUV instance directly affected by the fault. However, experimental results showed that in very few cases the propagation sequence generated by the CLP solver according to the proposed strategy, fails to propagate the corresponding fault when it is simulated on the faulty DUV.

In some cases, the CLP solver can be unable to find a propagation sequence starting from the provided configurations, because either the solver runs out of resources or the sequence does not exist. In both cases, a new activation sequence, and then a new pair of faulty and fault-free configurations, must be generated for the target fault.

Given a faulty and a fault-free configuration, to highlight that a propagation sequence does not exist, we define a constraint on DUV registers to assure that at least one register of the faulty DUV differ from the corresponding register of the fault-free DUV at each time step $T > 1$. On the contrary, the search is immediately stopped, and no solution is reported. Such a constraints avoid situations where the solver spends uselessly effort, as it cannot lead to the observability on POs if starting from different configurations, the faulty and fault-free DUVs evolve in the same configuration. In this context, the following constraint is required for the DUV of Figure 2:

`neg((REG[T] #= REG_F[T]) and (A[T] #= A_F[T])`
`and (B[T] #= B_F[T])).`

4.2 Search Procedures

Constraints described in the previous subsection are used to set up the problem of finding a propagation sequence as a CLP problem. Then, some search procedures, that exploit search strategies and heuristics, must be defined to force the solver to provide the solution (i.e., the set of values to be assigned to the logic variables for satisfying all the problem's constraints), when it exists. Therefore, we have defined the following predicate that exploits the *search/6* function of the ECLiPSe's IC library.

`search_func(A), search_func(B), search_func(IN1),...`

`search_func(L):-`
`    search(L,0,input_order,indomain,complete,[]).`

Such a function, whose signature is `search(Vars, Arg, Select, Choice, Method, Options)`, is a generic search routine which implements different partial search methods. It instantiates the variables `Vars` by performing a search based on the parameters provided by the user. In our case the search method performs a complete search routine which explores all alternative choices for each variable. The choice method `indomain` tries to find a solution by analyzing the variable values in increasing order, from the lower value in the variable range to the upper value. Experiments have been performed also with the choice method `indomain_random`, that uses the *built-in random/1* to create random numbers, but we observed that which such a method the number of backtrack steps is substantially increased.

The predicate `search_func` is called on each variable of the DUV. In this way, if a solution exists, the solver provides a value for each variable for each time step, thus generating the required propagation sequence.

4.3 Managing the CLP Complexity

Tools that exhaustively search for a solution of NP-hard problems frequently run out of resources when the state space to be analyzed is too large. The same happens for the CLP solver, when it is asked to find a propagation sequence on large sequential designs. To limit such a problem, heuristics is generally used for pruning the state space. However, this may prevent the solver from finding a solution (even if it exists), if the pruning is too restrictive. Thus, choosing a good heuristics is a very challenging task.

In this context, we have defined three strategies for managing the complexity of the CLP solver exploited by the fault-oriented engine. Two of them have been already presented in previous sections, however, for convenience of the reader, we summarize them here:

- The Tth element of arrays corresponding to states (and transitions) of the DUV are put in *xor* each other, to avoid that the solver wastes time to analyze configurations where more than one state variable is concurrently true. This drastically decreases the number of backtracking steps, especially for designs with many states and many registers. Consider that, in a design with n states and m registers, at each time step T, there are $\sum_{i=0}^{n-2} \binom{n}{n-i} * \prod_{j=1}^{m} 2^{sizeof(register_j)}$ configurations[2] where more than one state variable is concurrently true.

- A constraint on DUV registers is defined to assure that at least one register of the faulty DUV differ from the corresponding register of the fault-free DUV at each time step $T > 1$. On the contrary, the search is immediately stopped, and no solution is reported. Such a constraints avoid situations where the solver spends uselessly efforts, as it cannot lead to the observability on POs if starting from different configurations, the faulty and fault-free DUVs evolve in the same configuration.

A further strategy for managing the complexity of the CLP solver, that can be jointly used with the previous ones, consists of asking the solver to find a solution (i.e., a propagation sequence) starting with a small state space, that is incrementally enlarged until a solution is found (or execution time expires). Thus, the state space to be analyzed by the solver is restricted by limiting the range of the DUV

---

[2] *sizeof(register)* is for indicating the size of the register in bits.

PIs, similarly to what has been proposed for limiting the size of binary decision diagrams in the test generation strategy proposed in [20]. At the beginning, the ATPG statically fixes the values of all bits, but two, for each PI. In this way, only two bits can be changed by the CLP solver during the search, independently from the PIs range declared on the HDL description of the DUV. Then, the solver is asked to find a solution. If it fails, the ATPG opportunely increases the number of free bits of PIs. In particular, the ATPG engine searches for the constraints that induce the failure, and it frees the bits of the PIs involved in such constraints. Then, a new search session is launched. Such a process is iterated until a solution is found or execution time expires.

5. EXPERIMENTAL RESULTS

The efficiency of the CLP-based fault oriented ATPG has been evaluated by using the benchmarks described in Table 1, where the four left-most columns report the number of primary inputs (*PIs*), primary outputs (*POs*), flip-flops (*FFs*) and gates (*Gates*). Then, column *Trns.* shows the number of transitions of the S^2EFSM modeling the DUV and *GT (sec.)* the time required to automatically generate the S^2EFSM. Finally, Column *BC* reports the number of bit coverage faults injected into the designs.

The bit coverage fault model simulates under the following single fault assumptions:

Bit failures. Each occurrence of variables, constants, signals or ports is considered as a vector of bits. Each bit can be stuck-at zero or stuck-at one.

Condition failures. Each condition can be stuck-at true or stuck-at false, thus removing some execution paths in the faulty representation.

Bit coverage has been chosen since it is related to design errors [20] and it unifies into a single metrics the well known metrics concerning statements, branches and conditions coverage [21]. In addition, paths needed to activate and propagate faults from inputs to outputs of the design under verification (DUV) are also covered.

The considered benchmarks have been selected because they present different characteristics which allow us to analyze and confirm the effectiveness of the proposed approach. *b04* and *b10* have been selected from the well known ITC-99 benchmarks suite [22]. *b11m* is a modified version of *b11*, included in the same suite, created by introducing a delay on some paths to make it harder to be traversed. The original HDL descriptions of *b04*, *b10* and *b11m* contain a high number of nested conditions on signals and registers of different size. *bex*, *b00*, *b00z* and *root* contain conditional statements where one branch has probability $1 - \frac{1}{2^{32}}$ of being satisfied, while the other has probability $\frac{1}{2^{32}}$. Thus, they are very hard to be tested by random-based ATPGs. In particular, *bex*, *b00* and *b00z* are internal benchmarks, while *root* is a real industrial case, i.e., it is a module of a face recognition system.

The effectiveness of the CLP-based fault-oriented ATPG engine has been evaluated by applying the testing flow of Figure 1 where a genetic algorithm-based engine [2] has been used instead of the random-based engine for step 1, and the backjumping-based engine presented in [11] has been used for step 2.

Columns *Steps 1+2* of Table 2 report results achieved by applying step 1 and step 2 of the incremental test gen-

DUT	PIs	POs	FFs	Gates	Trns.	GT (sec.)	BC
bex	34	16	18	1131	10	0.1	378
b00	66	64	99	1692	7	0.1	1182
b04	13	8	66	650	20	0.3	408
b10	13	6	17	264	35	0.3	216
b11m	9	6	31	715	20	0.2	725
b00z	66	64	99	11874	9	0.2	1439
root	34	32	100	1475	10	0.2	1041

Table 1: Benchmarks properties.

eration flow of Figure 1. In particular, these columns show the achieved fault coverage (*FC%*), the number of faults observed but not detected, i.e., TBP faults (*TBP*), the number of faults not activated, i.e., TBA faults (*TBA*), the average length of the generated test sequences (*SLen*) and the test generation time (*T (sec.)*) by using the genetic algorithm-based and the transition-based engines in a cascade fashion. The fault coverage for *bex* is extremely low, since *bex* has been designed to be very difficult to be traversed, by imposing that: (i) after internal registers are initialized with data read from PIs, their values have a very high probability of being re-initialized moving to the next state; (ii) the majority of *bex* transitions cannot be fired if registers involved in the conditions of the corresponding enabling functions have been re-initialized. In this way, PIs values (and then the effect of faults) are difficult to be propagated towards POs.

Undetected faults have been classified as TBA or TBP, and activation sequences for TBP have been generated[3]. Then, the CLP-based fault-oriented engine has been applied to find propagation sequences for TBP faults. Columns *Prop.* and *T (sec.)*, below *Step 3*, reports, respectively, the number of TBP faults for which the CLP-based engine was able to generate a propagation sequence, and the corresponding execution time.

On the contrary, columns *Step 3 - pure CLP*, reports the results achieved by using the CLP-based engine without applying the strategies described in Section 4.3 for managing the CLP complexity. In this case, all TBP faults were aborted, since the CLP solver always run out of resources. This highlights the effectiveness of the strategies proposed for managing the CLP complexity.

Finally, the last three columns of the table report, respectively, the total fault coverage (*FC%*), the average length of test sequences (*SLen*) and the total generation time *T (sec.)* obtained by adopting all steps of the incremental testing flow shown in Figure 1.

Results show that the CLP-based engine, implementing the strategies described in Section 4.3, increased the fault coverage for all benchmarks without requiring long computation time. No fault has been aborted (i.e., the engine never run out of resources), even if some TBP faults remained untested, because no propagation sequence was found. The analysis of TBP faults not propagated highlighted the fact that many configurations allow TBP faults to be observed on internal registers (i.e., there exist many activation sequences), but very few of them allow TBP faults to be propagated. Moreover, such few configurations are difficult to be generated by using the engines adopted in step 1 and step 2, since they are not fault-oriented. To solve such a problem, in the future, the CLP-based fault-oriented engine will be

---

[3]Please, note that currently the CLP-based engine does not address TBA faults.

DUV	Steps 1 + 2					Step 3		Step 3 - CLP pure		Steps 1 + 2 + 3		
	FC%	TBP	TBA	SLen	T (sec.)	Prop.	T (sec.)	Prop.	T (sec.)	FC%	SLen	T (sec.)
bex	1.1	103	271	3	3.1	90	5.0	0	aborted	29.9	6	9.1
b00	52.5	64	498	3	2.9	84	2.5	0	aborted	59.6	7	5.9
b04	99.0	4	1	6	9.1	4	3.6	0	aborted	99.8	10	13.4
b10	94.0	13	0	11	6.8	12	3.3	0	aborted	99.5	18	10.1
b11m	23.6	117	124	59	16.3	313	36.7	0	aborted	66.8	71	56.7
b00z	13.8	131	497	6	12	613	9.1	0	aborted	56.4	30	18.5
root	84.0	63	82	42	5.2	22	6.0	0	aborted	86.1	80	11.2

Table 2: Experimental results.

extended for the generation of activation sequences too.

6. CONCLUDING REMARKS

In this paper, a CLP-based fault-oriented engine has been proposed together with a strategy for modeling EFSMs by means of constraint logic programming. To the best of our knowledge, this is the first paper addressing the problems of: (i) entirely modeling an EFSM by CLP; (ii) generating functional test pattern by combining the use of EFSM and CLP. Moreover, some strategies have been implemented to manage the CLP complexity, and experimental results showed that, in this way, the proposed engine is able to generate propagation sequences (thus, increasing the fault coverage) without running out of resources.

Currently, the engine is intended to be integrated in an incremental testing framework to generate propagation sequences for hard-to-detect faults observed on internal registers (i.e., activated), but not yet propagated towards POs. However, the capability of generating a propagation sequence for a fault depends on the quality of the starting EFSM configuration derived from the activation sequence. Thus, future works will be related to the extension of the CLP-based engine for generating activation sequences and then addressing TBA faults too.

7. REFERENCES

[1] F. Corno, G. Cumani, M. S. Reorda, and G. Squillero. *Effective Techniques for High-Level ATPG*. In *Proc. of IEEE ATS*, pp. 225–230. 2001.

[2] A. Fin and F. Fummi. *Genetic Algorithms: the Philosophers Stone or an Effective Solution for High-Level TPG?*. In *Proc. of IEEE HLDVT*, pp. 163–168. 2003.

[3] I. Ghosh and M. Fujita. *Automatic Test Pattern Generation for Functional Register-Transfer Level Circuits Using Assignment Decision Diagrams*. IEEE Trans. on Computer-Aided Design of Integrated Circuits and Systems, vol. 20(3):pp. 402–415, 2001.

[4] L. Zhang, I. Ghosh, and M. Hsiao. *Efficient Sequential ATPG for Functional RTL Circuits*. In *Proc. of IEEE ITC*, pp. 290–298. 2003.

[5] F. Xin, M. Ciesielski, and I. Harris. *Design validation of behavioral VHDL descriptions for arbitrary fault models*. In *Proc. of IEEE ETS*, pp. 156–161. 2005.

[6] J. Jaffar and M. J. Maher. *Constraint Logic Programming: A Survey*. Journal of Logic Programming, vol. 19/20:pp. 503–581, 1994.

[7] D. Lee and M. Yannakakis. *Online minimization of transition systems (extended abstract)*. In *Proc. of ACM STOC*, pp. 264–274. 1992.

[8] J. Li and W. Wong. *Automatic Test Generation from Communicating Extended Finite State Machine (CEFSM)-Based Models*. In *Proc. of IEEE ISORC*, pp. 181–185. 2002.

[9] M. Abramovici. *DOs and DON'Ts in Computing Fault Coverage*. In *Proc. of IEEE ITC*. 1993.

[10] F. Ferrandi, F. Fummi, and D. Sciuto. *Implicit test generation for behavioral VHDL models*. In *Proc. of IEEE ITC*, pp. 587–596. 1998.

[11] G. D. Guglielmo, F. Fummi, C. Marconcini, and G. Pravadelli. *Improving High-Level and Gate-Level Testing with FATE: a Functional ATPG Traversing Unstabilized EFSMs*. IEE Computers and Digital Techniques, vol. 1(3):pp. 187–196, 2007.

[12] S. Russel and P. Norvig. *Artificial Intelligence: A Modern Approach*. Prentice Hall, 2002.

[13] C. Pauli, M. L. Nivet, and J. F. Santucci. *Use of constraint solving in order to generate test vectors for behavioral validation*. In *Proc. of IEEE HLDVT*, pp. 15–20. 2000.

[14] R. Vemuri and R. Kalyanaraman. *Generation of design verification tests from behavioral VHDL programs using path enumeration and constraint programming*. IEEE Trans. Very Large Scale Integr. Syst., vol. 3(2):pp. 201–214, 1995.

[15] F. Ferrandi, M. Rendine, and D. Sciuto. *Functional Verification for SystemC Descriptions Using Constraint Solving*. In *Proc. of DATE*, pp. 744–751. 2002.

[16] F. Xin and I. G. Harris. *Test generation for hardware-software covalidation using non-linear programming*. In *Proc. of IEEE HLDVT*, pp. 175–180. 2002.

[17] D. Gajski, J. Zhu, and R. Domer. *Essential Issue in Codesign*. Thecnical report ICS-97-26, University of California, Irvine, 1997.

[18] G. Di Guglielmo, F. Fummi, C. Marconcini, and G. Pravadelli. *EFSM Manipulation to Increase High-Level ATPG Efficiency*. In *Proc. of IEEE ISQED*, pp. 57–62. 2006.

[19] M. Wallace and A. Veron. *Two problems-two solutions: one system-ECLIPSE*. In *IEE Colloquium on Advanced Software Technologies for Scheduling*, pp. 1–3. 1994.

[20] F. Ferrandi, F. Fummi, and D. Sciuto. *Test Generation and Testability Alternatives Exploration of Critical Algorithms for Embedded Applications*. IEEE Trans. on Computers, vol. 51(2):pp. 200–215, 2002.

[21] G. J. Myers. *The Art of Software Testing*. Wiley - Interscience, 1999.

[22] *High Time for High-Level Test Generation*. Panel at IEEE ITC, 1999.

AMS Verification
MTV 2007

Application of Automated Model Generation Techniques to Analog/Mixed-Signal Circuits*

Scott Little
School of Computing
University of Utah
Salt Lake City, UT, 84112 USA
Email: little@cs.utah.edu

Alper Sen
Freescale Semiconductor
Austin, TX 78729 USA
Email: alper.sen@freescale.com

Chris Myers
Electrical and Computer Engineering Department
University of Utah
Salt Lake City, UT, 84112 USA
Email: myers@ece.utah.edu

Abstract

Abstract models of analog/mixed-signal (AMS) circuits can be used for formal verification and system-level simulation. The difficulty of creating these models precludes their widespread use. This paper presents an automated method to generate abstract models appropriate for system-level simulation and formal verification. This method uses simulation traces and thresholds on the design variables to generate a piecewise-linear representation of the system. This piecewise-linear representation can be converted to a Verilog-AMS model or a Labeled Hybrid Petri Net formal model. Results are presented for the model generation, simulation, and verification of a PLL phase detector circuit.

1 Introduction

System-level design and verification challenges are increasing as analog/mixed-signal (AMS) designs become more common and functionally complex [1]. This increased complexity is leading to an increased number of functionality related bugs in AMS designs. These bugs are due in large part to inadequate modeling and verification methodologies for AMS design. Modeling is a challenge due to the need for models at differing levels of abstraction. Designers need and are comfortable with transistor-level models. However, abstract models are becoming necessary at the system level, but the expertise and time to create these models is often missing. Verification of AMS systems has traditionally been done in an ad hoc manner. More organization and formality is required to properly validate the increasing system complexity. It is clear that automated modeling and verification techniques can have a large impact in improving AMS system validation.

Abstract or reduced order modeling of linear systems is a well understood discipline with a strong foundation [7, 9]. This is not true for nonlinear systems in general. Although there are promising solutions for parts of the space [2, 3, 8]. There have been numerous investigations into the general problem although they have been met with scalability problems [10, 11]. As a result, automated abstract circuit modeling has not become common in AMS workflows.

Significant research effort has been invested in improving the uniformity and formality of design and verification of digital circuits. However, AMS verification has been slow to benefit from the efforts in the digital domain. Historically, digital verification has been done with switch-level simulation on very large designs. Comparatively, analog designs have been very small and designed using accurate transistor-level simulations at the block level. This history has resulted in two very different verification methodologies. The digital verification methodology is more formal and handled by a group of designers and verification engineers. Analog verification is ad hoc and often handled by a single designer. In the verification of a mixed-signal circuit, these two methodologies converge in a seemingly incompatible mix of methodologies.

---

*This research is supported by SRC contract 2005-TJ-1357 and an SRC Graduate Fellowship.

We believe that new verification methodologies are needed for mixed-signal designs. Based on this view, in our previous work we developed, LHPN Embedded/Mixed-signal Analyzer (LEMA), a tool to be used in an AMS verification workflow. LEMA is a formal verification tool for the verification of AMS systems. It has two main components, a model generation component and a verification component as shown in Figure 1. The verification component of the system is well described in previous work [4, 12–14]. LEMA is primarily intended to aid in AMS verification, but a difficulty with our previous work is creating an LHPN model. An initial discussion of the LHPN model generation component is found in [5]. This paper provides an extended discussion of the LHPN model generation as well as describing Verilog-AMS generation. Verilog-AMS model generation has been added to LEMA to provide designers with an automated method to obtain abstract models of their circuits. These generated models can be used for formal verification or system-level simulation.

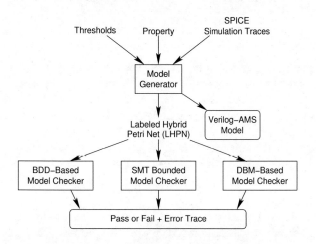

Figure 1. LEMA workflow diagram.

2 Motivating Example

The switched capacitor integrator circuit shown in Figure 2 is a circuit used as a component in many AMS circuits such as ADCs and DACs. Although only a small piece of these complex circuits, the switched capacitor integrator proves to be a useful example illustrating the type of problems that can be present in AMS circuit designs. Discrete-time integrators typically utilize switched capacitor circuits to accumulate charge. Capacitor mismatch can cause gain errors in integrators. Also, the CMOS switch elements in switched capacitor circuits inject charge when they transition from closed to open. This charge injection is difficult to control with any precision, and its voltage-dependent nature leads to circuits that have a weak signal-dependent behavior. This can cause integrators to have slightly different gains depending on their current state and input value. Circuits using integrators run the risk of the integrator saturating near a power supply rail. Therefore, the verification property to check for this circuit is whether the voltage V_{out} can rise above 2000mV or fall below -2000mV. It is essential to ensure that this never happens during operation under any possible permutation of component variations. For simplicity, we assume the major source of uncertainty in this example is that capacitor C_2 can vary dynamically by ± 10 percent from its nominal value. This circuit, therefore, must be verified for all values in this range [6].

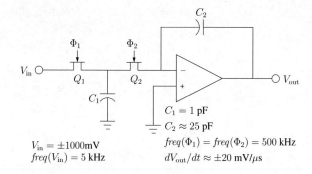

$V_{\text{in}} = \pm 1000\text{mV}$
$freq(V_{\text{in}}) = 5$ kHz
$C_1 = 1$ pF
$C_2 \approx 25$ pF
$freq(\Phi_1) = freq(\Phi_2) = 500$ kHz
$dV_{\text{out}}/dt \approx \pm 20$ mV/μs

Figure 2. A schematic of a switched capacitor integrator.

3 Model Generation

LEMA uses an ad hoc method to create a piecewise linear model for the system. The piecewise approximation for each variable is determined by the thresholds provided for the variable. The thresholds determine the separation points for each piece in the linearization. Based on the simulation traces a rate of change is calculated for each piece and transitions between the pieces are extracted. Model quality is dependent on the simulation traces, selected design variables, and thresholds.

LEMA's model generation algorithm consists of four major steps as illustrated in Algorithm 1. The inputs to genModel are a set of simulation traces, T; thresholds on the design variables, *thresholds*; a configuration file, c; and a property to verify, *prop*. The configuration file is used to customize model generation parameters. The property is introduced in the graph generation stage and only affects the LHPN output. The model generation is illustrated using simulation data from a $400\mu s$ transient simulation trace of the switched capacitor integrator using a capacitor value of 23 pF for C_2 shown in Table 1. The thresh-

Algorithm 1: genModel(T, $thresholds$, c, $prop$)
---
1 $G = \emptyset$;
2 **forall** $t \in T$ **do**
3      $binData = $ binData$(t, thresholds)$;
4      $rates = $ calculateRates$(t, c.window, binData)$;
5      $dmv = $ detectDMV$(t, c.\epsilon, c.ratio, c.\tau_{\min})$;
6      $dmv = $ calculateDMV(t, c, dmv);
7      $G = $ updateG$(G, binData, rates, dmv, prop)$;

olds on V_{in} and V_{out} are 0 V. The property for the design is $V_{\text{out}} \geq -2000 \wedge V_{\text{out}} < 2000$.

The first step, binData, sorts the data into bins according to thresholds. At $0\mu s$, V_{in} and V_{out} are below the threshold of 0V, so the bin encoding in the fourth column of Table 1 is 00 for $0\mu s$. The bin vector lists the bin values for V_{in} and V_{out} respectively. As V_{out} rises it exceeds the threshold of 0V resulting in a bin encoding of 01 as shown at the $46.98\mu s$ point of Table 1. The bins represent the individual pieces in the piecewise linear model. As a result, increasing the number of thresholds increases the accuracy of the model at a cost of increased complexity.

After bins have been assigned to each data point, the rates are calculated for each bin using calculateRates. A rate is calculated for each eligible data point. Not all points are eligible for rate calculations due to a low pass filtering technique used to smooth edge effects and transitory pulses. The low pass filtering uses a sliding window approach. The size of the window, $c.window$, can be configured in the configuration file. The sliding window approach works by calculating the rate between the current point and a point $c.window$ points further in time. For instance, in Table 1, the rate of V_{out} is calculated for 0.5 μs but not for $46.48\mu s$. A rate is not calculated for $46.48\mu s$ because the value of V_{out} $c.window$ points later is in a different bin.

In AMS designs, it is expected that digital signals are present. For efficiency purposes, digital-like signals are treated differently. Instead of allowing them to change with a specific rate, a constant value can be directly assigned to the variable at a specified time or when specific predicates become true. Discrete multi-valued (DMV) variables can be specified in the configuration file as well as automatically detected. A DMV variable is detected when it remains constant for a specified amount of time compared to the total time for the simulation. Remaining constant is defined as staying within an ϵ bound for at minimum time, $\tau_{\min}$. The DMV detection algorithm, detectDMV, is described in Algorithm 2. The algorithm tests each variable in the trace separately. The analysis begins with the first point and checks to see if the second point is equivalent within the specified ϵ. If equivalence is found the next point is tested. This occurs until a point is found, that is not equivalent. When this occurs, the time elapsed between the initial point and the current position is tested. If this time, $(\tau_{\text{pos}} - \tau_{\text{point}})$, is greater than $\tau_{\min}$, the value is added to the running total of constant time, t_{const}. The algorithm then continues from the next point. When all points have been analyzed, the ratio of constant time, τ_{const}, to total time for the trace, τ_{var}, is calculated. If this ratio exceeds the specified ratio, the variable is marked DMV.

Algorithm 2: detectDMV(t, ϵ, $ratio$, $\tau_{\min}$)
---
1 **forall** $var \in t$ **do**
2      $point, pos, \tau_{\text{const}} = 0$;
3   **while** $point < $ len$(t[var])$ **do**
4          $point = pos$;
5   **while** equiv$(\epsilon, point, pos + 1)$ **do**
6              $pos++$;
7   **if** $(\tau_{\text{pos}} - \tau_{\text{point}}) \geq \tau_{\min}$ **then**
8              $\tau_{\text{const}} += \tau_{\text{pos}} - \tau_{\text{point}}$;
9   **if** $(\tau_{\text{const}}/\tau_{\text{var}}) \geq ratio$ **then**
10         $dmv = dmv \cup var$;
11 **return** dmv;

The results of detectDMV are dependent on the type of variable. Input variables, specified in the configuration file, require extra calculation. The algorithm assumes that input DMV variables are periodic, so a minimum and maximum delay is calculated for each constant value. Constant values are calculated for all DMV variables. For instance, V_{in} is an input variable that is found to be a DMV variable. As can be deduced from column two of Table 1, V_{in} is found to have two constant values, $-1V$ and $1V$. Each value is found to have a delay between 99 and 101 μs as demonstrated in the final column of Table 1.

After the needed information has been calculated for a trace, updateGraph updates or creates the graph, G, with the new information. Nodes in the graph are created for each bin discovered as well as for each constant value of each DMV variable. The nodes are annotated with the appropriate rate, constant value, and delay information. Edges in the graph are created when a transition is observed between bins or constant values. The final graph is used to generate the Verilog-AMS and LHPN models.

3.1 Verilog-AMS model generation

Throughout the model generation process, ranges of values are extracted. Verilog-AMS does not support ranges of values. As a result, ranges are averaged for the Verilog-AMS model. A top level **inout** variable is created for each

Table 1. Partial simulation result with $C_2 = 23pF$ for the integrator.

Time (μs)	V_{in} (mV)	V_{out} (mV)	Bin	$\Delta V_{\text{in}}/\Delta t$ (mV/μs)	$\Delta V_{\text{out}}/\Delta t$ (mV/μs)	V_{in} time
0.00	-1000	-1000	00	-227.85	21.29	0.0
0.50	-1000	-999	00	0.0	21.74	0.5
⋮	⋮	⋮	⋮	⋮	⋮	⋮
46.48	-1000	-0.4	00	-	-	46.48
46.98	-1000	10	01	0.0	21.74	46.98
47.48	-1000	21	01	0.0	21.74	47.48
⋮	⋮	⋮	⋮	⋮	⋮	⋮
100.50	-1000	1174	01	-	-	100.50
100.54	-840	1174	01	-	-	100.54
100.62	-520	1176	01	-	-	100.70
100.78	120	1176	11	275.00	-21.08	0.08
101.00	1000	1174	11	0.0	-21.74	0.30
101.03	1.0	1173	11	0.0	-21.74	0.33
⋮	⋮	⋮	⋮	⋮	⋮	⋮
154.98	1000	0.3	11	-	-	54.28
155.48	1000	-11	10	0.0	-21.74	54.78
155.98	1000	-21	10	0.0	-21.74	55.28
⋮	⋮	⋮	⋮	⋮	⋮	⋮
200.00	1000	-978	10	-	-	99.30
200.04	840	-979	10	-	-	99.34
200.12	520	-980	10	-	-	99.50
200.28	-120	-981	00	-275.00	21.08	0.08
200.50	-1000	-978	00	0.0	21.74	0.30
200.53	-1000	-976	00	0.0	21.74	0.33
⋮	⋮	⋮	⋮	⋮	⋮	⋮
400.00	1000	-957	10	-	-	99.34

variable in the graph. `Vin_io` and `Vout_io` are the top level variables in the Verilog-AMS code for the switched capacitor integrator shown in Figure 3. This is the only information provided by the Verilog-AMS model for input variables. It is expected that the Verilog-AMS model will be driven by another circuit or test bench environment and this entity will provide the input. **real** variables of the form `<varname>_var` are created to hold the current value of each non-input variable, `Vout_var` for the switched capacitor integrator example. Variables with a rate (non-DMV variables) are also provided a **real** variable to store the current rate, for example `Vout_rate`.

The initial conditions for each variable and rate are set in the **initial_step** statement. Each edge containing a rate or constant value assignment is translated into a **cross** statement. The parameters for the cross statement are are extracted from the nodes between the edge. The source and sink nodes are compared. It is assumed that there is only one bin difference between the nodes. The variable for the changing bin is set as the compare variable. The threshold representing the division between the changing bins is the numerical value. The direction of the cross statement is calculated based on the signal's direction. If the signal is increasing, a 1 is used, and if the signal is decreasing, a -1 is used. In Figure 3, the first **cross** statement is created for an edge where V_{in} changes from $1V$ to $-1V$ by crossing the $0V$ threshold. As a result, `Vin_io` is the compare variable; `0.0` is the numerical value; and -1 is the direction. The sink place sets the rate of V_{out} to `0.020`, so this assignment is made to `Vout_rate` in the execution block of the **cross** statement. If there are multiple differences, an **if** statement will be used in place of the **cross** statement. The **if** statement is considered a less optimal solution as it has a weaker interaction with the simulator and does not provide a method to specify the direction of signal change. A global timer is added to update all rates at an appropriate interval which can be specified in the configuration file. For the switched capacitor integrator, an interval of $1\mu s$ is used to update the value of `Vout_var` based on `Vout_rate`. Finally, a **transition** statement is added for each non-input variable to quickly transition the value of the internal variable to the external interface.

```
module swCap(Vin_io,Vout_io);
  inout Vin_io, Vout_io;
  electrical Vin_io, Vout_io;
  real Vout_var, Vout_rate;
  analog begin
    @(initial_step) begin
      Vout_var = -1.00;
      Vout_rate = 0.020;
    end
    @(cross(V(Vin_io)-0.0,-1)) begin
      Vout_rate = 0.020;
    end
    @(cross(V(Vin_io)-0.0,1)) begin
      Vout_rate = -0.020;
    end
    @(timer(0.0,1e-06)) begin
      Vout_var = Vout_var + Vout_rate;
    end
    V(Vout_io) <+
      transition(Vout_var,1p,1p,1p);
  end
endmodule
```

Figure 3. Verilog-AMS code for the switched capacitor integrator.

3.2 LHPN model generation

The LHPN model generation is more direct than the Verilog-AMS because an LHPN is a directed graph that supports ranges on variables. LHPNs are not described in depth here, but a full description is found in [14]. In the translation, each node of the graph is translated to a place, p, in the LHPN. The fourth column of Table 1 shows four different bins discovered in the simulation trace. These four bins are represented by p_3 - p_6 of Figure 4a which models V_{out}. Figure 4b is the LHPN modeling V_{in}. This LHPN has two places, one for each constant value of V_{in}. Each edge in the graph is translated to a transition, t, in the LHPN. The annotations on each graph node are moved to the incoming transition. There are three types of annotations in Figure 4: predicates, assignments, and delay bounds. Predicates are specified with curly braces, {}; assignments are specified within angle brackets, ⟨⟩; and delay bounds are specified with square brackets, []. The input variables are modeled in the LHPN as a model for the environment is required to check properties of the system. The property is also integrated into the LHPN model. It is negated and added as an enabling condition to a single place and transition LHPN as shown in Figure 4c. When the transition fires indicating the property has been violated a special Boolean, *fail*, is set to true indicating an error to the analysis engine.

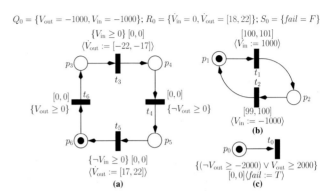

Figure 4. An LHPN model for the switched capacitor integrator.

4 PLL Phase Detector Example

Phase locked loops (PLLs) are notoriously difficult circuits to design and validate. There are a small number of major components to a PLL which traditionally include a phase detector, low pass filter, VCO, and a frequency divider. We analyze a phase detector as shown in the schematic diagram of Figure 5. The phase detector is used to measure the phase difference between two input signals and provide this information to the VCO. The VCO uses this information to adjust the phase of the internal PLL clock in order to align the phase of the two input signals. The inputs to the phase detector are *clk* and *gclk*. The $\overline{up}$ and $\overline{down}$ signals are asserted to provide instruction on how to adjust the VCO frequency.

The phase detector is simulated using a piecewise linear simulation input 1 μs long that represents reasonable clock skew for one input and a periodic clock signal for the other input. Simulations are also performed using two periodic signals of fixed, but different periods. These simulations are used to build the LHPN and Verilog-AMS models for the phase detector.

The Verilog-AMS and LHPN model are generated in approximately 20 seconds for the phase detector example using two 1 μs transient simulations of the piecewise linear input and a periodic clock input. Eight variables are required for model generation to accurately capture the state

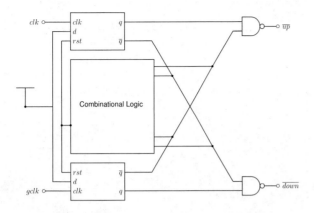

Figure 5. A schematic diagram for a PLL phase detector.

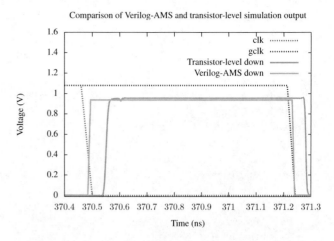

Figure 6. A comparison of a transistor-level and Verilog-AMS PLL phase detector models.

of the phase detector. Four of the signals are the inputs and outputs while the remaining four signals are chosen from within the internal signals of the latches. It is logical that signals from the state holding latches are required to delineate the states of the phase detector. As all eight variables are digital signals, each variable is assigned a single threshold equal to $\frac{V_{dd}}{2}$.

Comparison between simulation times for the transistor-level design and the Verilog-AMS model are performed using the same simulation inputs and simulator. Table 2 presents the results of these simulations. The first four simulations use one piecewise linear input and one periodic input. The final table entry is a result for two periodic inputs. A comparison of the waveforms produced by the two simulations is shown in Figure 6. There is a difference between the two waveforms, but the abstracted model is accurate enough to be used in a system-level simulation.

Table 2. A comparison of simulation times between the transistor-level model and the Verilog-AMS model.

Sim	Verilog-AMS (s)	Transistor (s)	Speed-up
0.5 μs	0.54	18.28	33.8
0.5 μs	0.54	17.92	33.2
1 μs	0.81	36.67	45.3
1 μs	0.81	40.46	49.9
2 μs	0.38	9.47	24.9

The LHPN model for the PLL phase detector is composed of 69 places and 87 transitions. The property $\neg(\overline{down} \wedge \overline{up})$ verifies in 0.3 seconds. This property is a sanity check on the outputs of the phase detector ensuring that $\overline{down}$ and $\overline{up}$ are not asserted at the same time. A more complex property for the PLL phase detector is shown as pseudocode in Figure 7. This pseudocode is a behavioral description of the correct input/output operation of the phase detector. This property can not be specified directly in LEMA's property language. To verify the property it is necessary to convert the property to an LHPN with the appropriate *fail* transitions. The resulting property LHPN is composed of 20 transitions and 14 places. This property verifies in 2.12 seconds.

```
if gclk 1 → 0 then
   if up = 1 then up = 0 within 5 ns
   elsif down = 0 then down = 1 within 5
ns
if clk 1 → 0 then
   if down = 1 then down = 0 within 5 ns
   elsif up = 0 then up = 1 within 5 ns
```

Figure 7. A property to verify for the PLL phase detector.

5 Conclusions

Algorithms have been developed to generate Verilog-AMS and LHPN models from a set of simulation traces. These algorithms create a piecewise linear model for the circuit. These methods are evaluated using a PLL phase detector circuit. The methods are successful in the case, but during the course of applying the methodology to other circuits several lessons are learned. Modeling quality is dependent on a number of factors which are difficult to understand and control for many circuits. To create a quality model the number and type of simulations must adequately characterize all behaviors of the circuit. It may be difficult to determine and create the needed simulations. The proper variables to distinguish individual states must be present in the simulation traces. Even with a digital circuit like the PLL phase detector it is initially non-obvious which signals are required to delineate the state of the system. This problem is even more complex when dealing with a circuit containing more analog components. The modeling method does not handle start-up conditions well as it is not often possible to find a state variable that distinguishes the start-up of the circuit from steady state operation. The results of the phase detector do show the promise of abstract modeling to decrease system-level simulation. Further exploration in automated abstract model generation should be undertaken.

References

[1] H. Chang and K. Kundert. Verification of complex analog and RF IC designs. In *Proceedings of the IEEE*, volume 95, pages 622–639, Mar. 2007.

[2] X. Lai and J. Roychowdhury. Tp-ppv: Piecewise nonlinear, time-shifted oscillator macromodel extraction for fast, accurate pll simulation. In S. Hassoun, editor, *Proc. International Conference on Computer Aided Design (ICCAD)*, pages 269–274. ACM Press, 2006.

[3] P. Li and L. T. Pileggi. Norm: compact model order reduction of weakly nonlinear systems. In *Proc. Design Automation Conference (DAC)*, pages 472–477, 2003.

[4] S. Little, N. Seegmiller, D. Walter, C. Myers, and T. Yoneda. Verification of analog/mixed-signal circuits using labeled hybrid petri nets. In *Proc. International Conference on Computer Aided Design (ICCAD)*, pages 275–282. IEEE Computer Society Press, 2006.

[5] S. Little, D. Walter, and C. Myers. Analog/mixed-signal circuit verification using models generated from simulation traces. In *Automated Technology for Verification and Analysis (ATVA)*, Lecture Notes in Computer Science. Springer, 2007.

[6] C. J. Myers, R. R. Harrison, D. Walter, N. Seegmiller, and S. Little. The case for analog circuit verification. *Electronic Notes Theoretical Computer Science*, 153(3):53–63, 2006.

[7] A. Odabasioglu, M. Celik, and L. T. Pileggi. Prima: passive reduced-order interconnect macromodeling algorithm. *IEEE Transactions on Computer-Aided Design of Integrated Circuits and Systems*, 17(8):645–654, 1998.

[8] J. R. Phillips. Projection-based approaches for model reduction of weakly nonlinear, time-varying systems. *IEEE Transactions on Computer-Aided Design of Integrated Circuits and Systems*, 22(2):171–187, 2003.

[9] L. T. Pillage and R. A. Rohrer. Asymptotic waveform evaluation for timing analysis. *IEEE Transactions on Computer-Aided Design of Integrated Circuits and Systems*, 9(4):352–366, 1990.

[10] M. Rewieński and J. White. Model order reduction for nonlinear dynamical systems based on trajectory piecewise-linear approximations. *Linear Algebra and its Applications*, 415(2–3):426–454, jun 2006.

[11] S. K. Tiwary and R. A. Rutenbar. Faster, parametric trajectory-based macromodels via localized linear reductions. In S. Hassoun, editor, *Proc. International Conference on Computer Aided Design (ICCAD)*, pages 876–883. ACM Press, 2006.

[12] D. Walter, S. Little, and C. Myers. Bounded model checking of analog and mixed-signal circuits using an smt solver. In *Automated Technology for Verification and Analysis (ATVA)*, Lecture Notes in Computer Science. Springer, 2007.

[13] D. Walter, S. Little, N. Seegmiller, C. J. Myers, and T. Yoneda. Symbolic model checking of analog/mixed-signal circuits. In *Proc. of Asia and South Pacific Design Automation Conference (ASPDAC)*, pages 316–323, 2007.

[14] D. C. Walter. *Verification of analog and mixed-signal circuits using symbolic methods*. PhD thesis, University of Utah, May 2007.

Functional Modeling and Testbenches
MTV 2007

An ADL for Functional Specification of IA32

Wei Qin
Boston University
Boston, MA 02215
wqin@bu.edu

Asa Ben-Tzur   Boris Gutkovich
Intel Corporation
Haifa, Israel
{asa.ben-tzur,boris.gutkovich}@intel.com

Abstract

Many architecture description languages (ADL) have been recently proposed to automate the design of new microprocessors and their related development tools. However, none of those comes close to fully describing the IA32 architecture. In this paper, we present an ADL that is custom designed for the IA32 architecture. The ADL supports the unique features of IA32 that are generally ignored by other ADLs. It features a high-level type system, simple syntax, and a well-understood computation model. The ADL is analyzable in that it preserves high-level architectural features in its descriptions. It is also executable since it has bit-accurate semantics free of ambiguity. The ADL is expected to be used as a unified IA32 description for an instruction set simulator, a functional test generator, and possibly other tools.

1 Introduction

Many architecture description languages (ADL's) have been recently proposed to automate the development of new architectures. Their primary application has been to configure the development tools such as software compilers, instruction set simulators (ISS's), and sometimes hardware synthesizers, so that these tools can quickly adapt to the choices of the designer. An important objective of most ADLs is the scope of coverage: a wider range of architecture is always preferable over a narrower one. However, despite many years of research, it still seems an extremely challenging task to create a universally flexible ADL for all known architectures and all development tools. This is largely due to the diversity of existing processor architectures, and the radically different requirements imposed by different development tools. Among the known architectures, CISC ones such as the IA32 [10] pose the biggest challenge to ADL designers. As for the development tools, compilers are usually the toughest challenge since they require human intelligence to handle tasks such as simplifying architectural features and translating them to data models usable by optimization algorithms.

This paper presents a new ADL in progress with a different objective from the aforementioned coverage objective. The ADL focuses on one mainstream architecture, the IA32. Due to its narrow focus, this work can afford to cover the special IA32 architectural aspects that previous ADLs neglected. For example, the operating modes, the virtual memory model, the exception model, the use of prefixes, and the size attributes for operands. The initial goal of the ADL is to provide an analyzable and executable architectural specification for a functional test generator, and an ISS. After the accomplishment of the initial goal, the scope of the ADL will be extended to cover the microarchitecture, and to support other development tools.

The paper is organized as follows. Section 2 briefly reviews related work in the area of architecture description. In Section 3, we present the details of the proposed ADL. We will emphasize on the unique features that are specifically created for this ADL. We summarize the contributions and discuss some important design decisions in Section 4. In the end, Section 5 concludes the paper and points out the ongoing work.

2 Related Work

Traditionally, the term ADL has been used for not only description languages that cover instruction set architectures, but also those that cover microarchitectures. Depending on the nature of included information, ADLs have been categorized into behavioral ADLs, structural ADLs, and mixed ADLs. Behavioral ADLs cover only instruction set information. They include ISPS [3], nML [5], and ISDL [8]. Structural ADLs resemble hardware description languages (HDLs), and are used for describing microarchitecture information. They include MIMOLA [16] and UDL/I [1]. Mixed ADLs contain both instruction-set architecture (ISA) and microarchitecture information. They adopt various custom schemes to separate the description of these two types of information. Example ADLs in this category include

LISA [13], EXPRESSION [9], MADL [14], etc. Most of these ADLs are designed to support relatively regular and simple architectures.

Some related work is not generally recognized as ADLs, but nevertheless provides some ISA information in a less self-contained manner. For example, the LISP-like code translation pattern used in the backend of GCC [6] is one special form of architecture description, though it is only useful for the purpose of code generation. The binary encoding information supported by SLED [15] is a subset of ISAs. Compared to real ADLs, these description schemes serve a narrower purpose and thus can often achieve excellent description quality for mainstream architectures. However, due to the incompleteness of information, their usage is limited. They are better viewed as configuration systems for their associated tools.

An important objective of the previous description languages and systems is the coverage of the architecture space. In all cases, the purpose of having a description system is to support the associated retargetable development tools. It is necessary that the description system achieve a coverage scope that is no smaller than those of its development tools. However, since every real-world architecture has some unique features, and new features continue to be introduced in new processors, it is extremely difficult to design a formal ADL that covers all features of all architectures. The coverage problem is especially difficult for CISC architectures and irregular DSP architectures. For example, among the existing ADLs, none has claimed a nearly complete coverage of the special features of the IA32 architecture, such as the size attribute associated with code segments and its complex memory views. In real life, practical hacks are often used to complement the structured description systems. For example, in GCC, C code hacks are used to complement the descriptive code patterns. One clear drawback of the hacks is that their information cannot be analyzed by the ADL compiler, and cannot be reused by other development tools. Therefore users will have to duplicate their information in different forms for different tools.

3 Proposed ADL

3.1 Motivation

The proposed ADL belongs to the behavioral ADL category. Instead of focusing on general and flexible coverage, the goal of the ADL is to enable an analyzable and executable specification of the IA32 architecture. An analyzable specification allows the associated functional test generator to extract the relevant properties of instructions. It requires that the ADL preserve in a descriptive way all high level architectural information such as data structures, memory address translation schemes, instruction semantics, and exceptions. An executable specification is useful for the ISS. It requires that the description be based upon bit-accurate semantics without ambiguity.

The narrow architecture scope of the work differs from all previous ADLs. It is justified by the following reasons.

First, IA32 is the dominant architecture that generates the most revenue of all microprocessors. Given the large amount of engineering effort that has been continuously invested, it is economically worthwhile to create a dedicated ADL for it. Successful outcome of the work will benefit all future engineering work around the architecture.

Second, IA32 has many special features that are overlooked by the existing ADLs. In the reviewed systems in Section 2, only GCC and SLED have limited support of IA32 for specific tools. Therefore, it is necessary to study new description techniques to address the following unsolved technical challenges.

- Various modes of operation,
- Sophisticated addressing modes,
- Switchable 16-bit and 32-bit operand/address sizes,
- Complex instruction format with prefixes,
- Complex instruction behaviors including SIMD,
- Three memory views, paging, and segmentation,
- Side-effects and exceptions.

We feel it necessary to first narrow the architecture scope to solve the above challenges, before moving toward a wider scope.

Lastly, many development tools separately encoded the architecture information in their respective forms. This creates verification and maintenance nightmares. An ADL providing a unified architecture specification for all tools is highly desirable.

3.2 Organization of ADL

Like all existing behavioral ADLs, the proposed ADL is instruction-centric. It specifies the properties of instructions in a well-structured and concise manner. Syntactically, an ADL description is an unordered list of declarations. This section presents the various declarations. Emphasis will be placed on those features that are unique to the proposed ADL. The other features will be briefly described.

3.2.1 Preprocessing Declarations

Similar to many high level languages, some preprocessing capability has been built into the ADL for convenience of use. This contains the *include* and the *alias* declarations. An *include* declaration is similar to its namesake preprocessing command in C. Its use allows an architecture description to be divided into multiple files for easy editing and managing. An *alias* declaration is similar to the macros used in C. It is typically used to assign symbolic names to long operands.

3.2.2 Data-type Declarations

The proposed ADL defines a type system that supports both scalar and aggregate types. Scalar types include Boolean, enumeration, and fixed width integer types. Among those, the fixed-width integer type is most flexible. It can be either a basic integer type, or a complex type with subfields. The subfields must have narrower fixed-width integer types, and may overlap with each other. The use of subfields makes it convenient to describe the complex data structures in IA32, such as segment descriptors. It is permissible to denote a subfield as a reserved field so that any attempt to use it individually will receive a warning. It is also possible to assign a constant value to a reserved subfield.

The example below shows a 64-bit unsigned integer type that defines the segment descriptor structure of IA32. The "base" and "seg_limit" fields are made up of non-contiguous bits.

```
typedef uint<64> {
    base     : uint<24> = {63:56} {39:16},
    seg_limit : uint<20> = {51:48} {15:0},
    g        : uint<1>  = {55},
    d_b      : uint<1>  = {54},
    l        : uint<1>  = {53},
    avl      : uint<1>  = {52},
    p        : uint<1>  = {47},
    dpl      : uint<2>  = {46:45},
    s        : uint<1>  = {44},
    type     : uint<4>  = {43:40},
} seg_des_t ;
```

Additionally, type inheritance is supported for fixed-width integer types. It allows a general data type to be refined to a more specific data type. For example, the above segment descriptor "seg_des_t" can be refined to the following data segment descriptor.

```
typedef seg_des_t {
    s = 0,
    t = 0 : uint<1> = {43},
    e     : uint<1> = {42},
    w     : uint<1> = {41},
    a     : uint<1> = {40},
} data_seg_des_t;
```

The data segment descriptor type above is a segment descriptor. It inherits all subfields of the "seg_des_t" type. But it also has its own subfields *t*, *e*, *w*, and *a*. It is characterized by the signature values of the *s* and *t* bits.

Aggregate types includes tuple and array. The tuple type is especially useful for specifying the logical memory address of IA32, which contains a segment selector and an offset.

In summary, the type system in the proposed ADL borrows some features from high level programming languages. It is convenient to use, and is helpful to preserve the high level architectural information that is required by tools such as the functional test generator. Such a comprehensive type system does not exist in previous ADLs.

3.2.3 Instruction Description Scheme

The description scheme contains three layers: physical storages, operands, and instructions. The physical storage declarations define the physical registers, and the memory (discussed in Section 3.2.5). The declaration of a physical register can utilize the fixed-width integer data type. For example, integer registers of IA32 can utilize the following type:

```
typedef uint<32> {
    half   : uint<16> = {15:0},
    hquart : uint<8>  = {15:8},
    lquart : uint<8>  = {7:0},
} ireg_t;

register iregs[8]:ireg_t;
```

The operand declarations form the logical operands based on physical storages, for example, 8-bit register operands. The operand declarations are not essential but they serve as a convenient layer to group common portions of instructions. Their use greatly reduces the length of the overall specification. For each operand, its assembly syntax, binary coding, and semantics need to be defined.

The "ireg_t" type in the previous register example is a 32-bit integer type for 32-bit registers. Its subfields can be utilized to refer to the narrower register operands of IA32. For example, the AL operand can be referred to as "iregs[0].lquart", and the overlapping AX operand can be referred to as "iregs[0].half". All 8-bit register operands can be grouped as the following operand.

```
operand r8 {    /* 8 bit registers */
   option syntax = "AL", coding = 000,
          semantics = iregs[0].lquart;
   option syntax = "CL", coding = 001,
          semantics = iregs[1].lquart;
   option syntax = "DL", coding = 010,
          semantics = iregs[2].lquart;
   option syntax = "BL", coding = 011,
          semantics = iregs[3].lquart;
   option syntax = "AH", coding = 100,
          semantics = iregs[0].hquart;
   option syntax = "CH", coding = 101,
          semantics = iregs[1].hquart;
   option syntax = "DH", coding = 110,
          semantics = iregs[2].hquart;
   option syntax = "BH", coding = 111,
          semantics = iregs[3].hquart;
}
```

The instruction declarations are based on the physical storages and the operands. They enumerate the list of instructions in the architecture reference manual. For each instruction, its binary encoding, assembly syntax, and semantics are specified. A set of operators are defined for specifying instruction semantics. In addition, functions can be defined and referenced to improve the modularity of semantics description. The example below specifies a simple *mov* instruction that copies data from the "Sreg" operand to the "RM16" operand.

```
instruction mov_sreg (RM16, Sreg) {
    syntax = "MOV $1, $2",
    coding = 10001010 $1.$1 $2 $1.$2,
    semantics = { $1 = $2;}
}
```

Overall, the instruction description scheme of the proposed ADL uses a similar structure to the well-known nML [5]. The storage, operand and instruction layers form the AND-OR tree structure first introduced in nML. The structure factorizes common parts of similar instructions into the operand layer. An instruction declaration is the root of an AND-OR tree, its operands form the intermediate nodes, and the physical storages are the leaf nodes. A single tree represents a factorized set of similar instructions. The ADL compiler will expand the tree to obtain the full specification of the set of instructions. Given the fact that it is not a new feature introduced by the proposed ADL, we will save the explanation of the concept. Interested readers can refer to [5] for more details.

3.2.4 Environment

IA32 is a very sophisticated ISA. Its many features cannot be modeled by previously reported ADL techniques, for example, the operating modes of the processor that enable backward compatibility, and the size attribute of the code segment descriptor that affects the behavior of most instructions in the segment. But these global properties do not belong to any instruction, their effect on instructions cannot be described in an instruction-centric ADL.

To solve the problem, the proposed ADL introduces the notion of environment. An environment represents a global state that is not a part of an instruction but affects its behavior. Such states include the architecture generation, the operating mode, and the size attribute of the code segment.

In addition to global states, environments are used to represent instruction prefixes. Instruction prefixes are optional parts of an IA32 instruction. Some prefixes interfere with the global states such as the size attribute of the code segment. One cannot interpret the semantics of these prefixes out of the context of the global states. For example, the group-4 prefix toggles the size attribute defined by the code segment environment. So the same prefixed instruction may show different behaviors when placed in different code segments. To conveniently describe such interference with global states, we treat prefixes as environments too.

An environment is defined by the keyword *environment*, followed by its name, and its enumeration type. Below is one example environment "seg_size" and its type.

```
typedef enum {
    _16_bit,
    _32_bit,
} seg_size_t;

environment seg_size : seg_size_t;
```

The above environment is related to the operand/address size of a code segment. To use an environment, one can include the environment statements (#if, #elif, #else, #endif) in a description. In this example, if the size attribute of the code segment is 32-bit, or if the size is 16-bit but a group-4 prefix (another environment) overrides the segment size attribute, the instruction will use 32-bit operands. Otherwise, the instruction should use 16-bit operands. This conditional situation is illustrated in the example below.

```
#if (seg_size==_32_bit && !defined(pre_g4))
  || (seg_size==_16_bit && defined(pre_g4))
    // define 32 bit behavior here
#else
    // define 16 bit behavior here
#endif
```

Although the syntax of the environment statements is similar to some preprocessing commands of C/C++, their meanings are different. In C/C++, the preprocessing conditions are static and resolved prior to compilation. In the proposed ADL, the conditions are not always static. For example, the segment size of an instruction is unknown at the time when an ADL description is compiled. Therefore, all combinations of the environmental conditions are preserved by the ADL compiler in a database. When the database is queried for the details of an instruction, proper environments must be supplied. The environmental condition is then resolved and the corresponding descriptions of the instructions are returned.

3.2.5 Memory Modeling

The IA32 architecture has several memory views under the protected mode: a physical view, a linear view and a logical view. Conceptually, a logical memory access is mapped to a linear memory access via segmentation, and a linear memory access is mapped to the physical memory via paging. Though seemingly redundant, having all three views is necessary to preserve backward compatibility of the architecture.

Some tools, such as the functional test generator, need to understand the mapping algorithms and the exceptions that they may cause. However, some other tools, such as the compiler, may not need the details of such mapping. Therefore, we designed a scheme that allows the mapping details to be separated from the regular instruction semantics. This is done through the memory declaration, which specifies the memory views and their relationships. For example, under an environmental condition where both paging and segmentation are enabled, one can specify the memory views in the following way.

```
/* actual memory, this will create the
   storage */
memory physical_mem : physical_addr_t;
```

```
/* linear view of the memory */
memory linear_mem : linear_addr_t = {
    target  = physical_mem,
    mapping = paging;
}

/* logic view of the memory */
memory logical_mem : logical_addr_t = {
    step    = (0, 1);
    target  = linear_mem,
    mapping = logical_to_linear;
}
```

The memory keyword specifies a view of the memory. Inside the memory declaration, one can define its address type, the target memory view, and the name of the function that maps its addresses to the target addresses. In the above specification, the linear memory maps to the physical memory via the function "paging", and the logical memory maps to the linear memory via the function "logical_to_linear". Both "paging" and "logical_to_linear" are user-defined mapping functions in the ADL description. A mapping function always takes two arguments: the address and the access type (read or write). When the ADL descriptions access a memory address using the indexing operator [ ], e.g. "linear_mem[100]", the address (100 in this example) is passed to the mapping function as the first argument. The ADL compiler will supply the second argument depending on the context of access, i.e. whether the "linear_mem[100]" operand appears on the left side or right side of the assignment operator. The access type is useful to check access permissions in the protected mode. Within the body of a memory declaration there can also be an optional attribute named "step", which defines the increment step of the byte address. This is necessary when the address is a tuple because the ADL does not have the knowledge of arithmetic operations on tuples.

The example below is the simplified "logical_to_linear" function. The function translates a logical address to a linear address. It first checks the *ti* field of the segment selector to decide which descriptor table to use. Then it looks up the corresponding descriptor table (pointed to by either GDTR.base or LDTR.base) in the linear memory, gets the base address of the segment descriptor, and finally computes the linear address.

```
function logical_to_linear
    (logical_addr_t la,
     access_t ac) :
    linear_addr_t
{
    var seg_sel_t sel;
    var seg_des_t sd;
    var uint<32> des_tab;

    sel = la.$1;
    if (sel.ti == 0b0)
        des_tab = sel.ti==0b0 ?
                    GDTR.base : LDTR.base;

    /* look up LDT or GDT first */

    sd = linear_mem[<seg_des_t>des_tab +
                    la.$1 .index];

    return sd.base + la.$2;

    /* exceptions omitted here,
       see following section */
}
```

To obtain the descriptor, the above mapping function looks up the descriptor table at *des_tab* in the linear memory. The lookup triggers the other memory mapping function "paging". The data type and size of the access is determined by the context. In this case, the type is the segment descriptor.

When a mapping function is evaluated, it may trigger exceptions. Exceptions are omitted in the above example for simplicity. They will be explained in the following section.

3.2.6 Exception and side effect

Exceptions and side-effects are important parts of instruction semantics. They have been either ignored by existing ADLs, or merged into the description of regular semantics. Merging them into regular semantics is not desirable because of the following reasons:

- Most architecture reference manual separates them from the description of regular instruction semantics. It involves non-trivial effort to merge them.

- It requires extra analysis to separate them from regular instruction semantics when a tool needs to treat them differently. The analysis may be very difficult.

To avoid such unnecessary merging and separation, the proposed ADL separates them from regular instruction semantics. In our ADL, a special label "side_effect" is reserved for highlighting exceptions and other side effects such as changes of machine flags. Below we list two example functions demonstrating the description of those. The first one is the previously omitted portion of the memory mapping function "logical_to_linear". It raises exceptions when the segment is not valid, or when the segment is non-writable. The second function performs 16-bit addition and modifies several machine flags.

```
function logical_to_linear
    (logical_addr_t la,
     access_t ac) :
    linear_addr_t
{
/* regular part omitted */
......

side_effect:

#if mode==protected

    /* segment not present */
    if (sd.p==0) raise(NP, 0);
```

```
        /* write to non-writable segment */
        if (sd isa code_seg_des_t ||
            (sd isa data_seg_des_t && ac==_W &&
             sd.type{1}==0))
            raise(GP, 0); /* raise exception */

/* more exceptions ...... */

#endif // mode==protected

}

function fadd16
    (uint<16> oprnd1,
     uint<16> oprnd2) : uint<16>
{
    uint<16> result;

    result = oprnd1 + oprnd2;
    return result;

side_effect:
    /* carry, sign, overflow, parity,
       auxiliary carry */
    CF = result < oprnd1;
    SF = result{15};
    OF = (oprnd1 ^ oprnd2 ^ 0xFFFF) &
         (oprnd1 ^ result);
    PF = !(result{0} ^ result{1} ^
           result{2} ^ result{3} ^
           result{4} ^ result{5} ^
           result{6} ^ result{7});
    AF = (oprnd1{3:0} + oprnd2{3:0}) >> 4;
}
```

The "side_effect" label provides a separation of normal semantic actions and side effects. When necessary, more labels may be introduced to separate semantic actions of different nature in the future. Marking the nature of semantic actions is the only role of labels. They cannot be used as jump targets since there exists no equivalent of the "goto" statement in the ADL. A "goto" poses unnecessary challenges to the ADL compiler due to the complex control flow it brings.

3.2.7 The computation model

The ADL provides a functional description of the IA32 architecture. The description is instruction-centric. Its core part is the specification of instruction semantics. In order to provide analyzable instruction properties to tools such as the functional test generator, it is necessary to base the semantics description on a descriptive and flexible computation model.

In general, data flow models are suitable to represent the semantic statement lists. In the proposed ADL, we chose a well-studied data-flow model named Boolean Dataflow (BDF) [7] as the underlying computation model. It is sufficient to represent all the semantic statements in our ADL. Compared to simpler models such as the Synchronous Dataflow (SDF) [12] or the primitive data-dependency graph, BDF is more flexible and supports the modeling of control flow. Compared to more complex models such as the Dynamic Dataflow (DDF) [4], Kahn Process Network (KPN) [11] or the many variants of Control and Dataflow Graph (CDFG) [2], BDF is more amenable to formal analysis.

The BDF model is first introduced in [7]. It was later studied in the Ptolemy framework by Joseph Buck [4]. A BDF is a token flow graph. It contains four types of nodes:

- Operator nodes – regular computation nodes
- IO nodes – for input and output
- Branch nodes – de-multiplexer
- Merge nodes – multiplexer

The nodes are connected by directed edges. They consume and produce valued tokens, which are buffered on the edges. The state of a BDF is characterized by the tokens on its edges. The word "Boolean" in the name of the model means that the consumption/production rates are binary, i.e. when a node fires, each of its incoming edges may consume 0 or 1 token, and each outgoing edge may produce 0 or 1 token.

Operator nodes are the simplest. Each firing of an operator node always consumes one token from each incoming edge and places one token to each outgoing edge. The IO nodes have similar firing rules as the operator nodes. But an input node may only have outgoing edges, and an output node may only have incoming edges. Figure 1 shows the state of an example operator node before and after its firing. A firing is generally considered an atomic action.

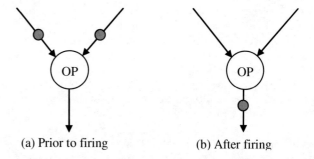

(a) Prior to firing   (b) After firing

Figure 1. Operation node example

A branch node has two incoming edges – a data edge and a control edge – and a list of output edges. A firing of a branch node consumes one token on each input edge. Depending on the value of the control token, the data token is forwarded to one of its output edges. A merge token is the opposite of a branch. It has one control input, a list of data inputs, and one data output. Depending on the value of the control token, one input is forwarded to the output edge. Figure 2 shows a branch example and a merge example.

The BDF model has precise semantics and is executable. The semantic expression and statements in the ADL can all be translated into BDFs. Figure 3 demonstrates the translation result for an instruction with the semantics $EAX = EAX + logical\_memory[(DS, SI + 12)]$.

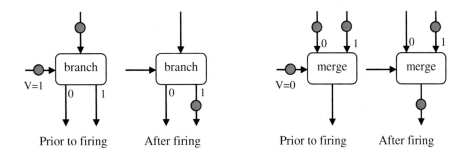

Figure 2. Branch and merge nodes example

The BDF used in the ADL supports hierarchical composition. For example, the logical memory lookup node in Figure 3 can be further expanded into Figure 4, which includes the function "logical_to_linear" in "ia32.adl". The BDF in Figure 4 receives the logical address as its input. It disassembles the address into the segment selector and the offset. Depending on the subfield "ti" of the selector, it chooses to look up either the GDTR or the LDTR. The chosen descriptor table is then indexed by the "index" field of the segment selector. The result is a segment descriptor, whose "base" subfield is added to the offset element of the logical address to form the linear address. Finally, the linear memory is looked up using the linear address.

The linear memory node can be further expanded. Apparently, the expansion provides more details but also leads to a more complex BDF. It is up to the tool to decide whether or not to query the ADL framework for such an expansion. When the view of the logical memory is sufficient for a task, the expansion can be omitted.

Side-effect statements can also be translated and attached onto the main BDF. When multiple exceptions are added to the main BDF, their priorities need to be determined. The proposed ADL defines the priorities of exceptions according to the firing order of the main BDF nodes that they depends on. For example, if an exception depends on the main BDF node A to fire, and another exception depends on the main BDF node C. And if A dominates C, i.e. A always fires prior to C, the first exception has a higher priority. However, if the firing order of A and C cannot be decided statically, or if the reference manual explicitly states that the second exception should have higher priority, then a dependency between the first exception on C should be given in the description. A condition can be used to suppress the first exception when the second exception should indeed fire.

As explained in [4], BDF can also express loops. Therefore, the model is general enough to model the string instructions of IA32. The ADL supports *while* loop statements in instruction semantics descriptions for string instructions.

4 Discussions

In summary, the proposed ADL is the first that fully targets the IA32 architecture. It made the following unique contributions to the ADL field:

- It supports a comprehensive type system that resembles high level programming languages. The type system helps to preserve and to expose high-level architectural information to the tools that use the ADL. Previous ADLs do not have such sophisticated type system.

- It supports the concept of environment. Environments are useful to model special IA32 features such as the architecture revisions, the operating modes, and the prefixes. Previous ADLs do not have similar features.

- It supports the description of different memory views and their relationships. Previous ADLs exclude the description of memory mapping.

- It supports explicit description of exceptions and side-effects. Previous ADLs either ignore these, or merge them into the regular semantics.

- It uses the well-understood BDF model to represent behavioral semantics of instructions. The model is analyzable, and sufficiently flexible for all semantic statements in the ADL. Previous behavioral ADLs are based on pure dataflow model and lack the flexibility for specifying loops.

It should be pointed out that the ADL is neither intended as a replacement of the architecture reference manual, nor a complete specification of the architecture. It focuses on the description of individual instructions, including their semantics, assembly syntax and binary encoding. Architectural level semantics, such as the sequential fetch-decode-execution semantics, the automatic increment of the program counter, the precise exception semantics, the endianness of memory access, and the meanings of the environments, are not explicitly specified in the ADL. Instead, they remain as underlying assumptions of the ADL and the supported development tools. As a matter of fact, there is no obvious advantage of describing those less-structured architectural aspects in an ADL. Including those will clutter the syntax of the ADL, and add no practical productivity benefit.

Currently, the ADL is still in progress. Ongoing work includes floating point support, and a comprehensive description of all IA32 instructions. We expect a significant amount of effort to be spent on verifying the description. An instruction-set simulator generated from the ADL description is helpful in verifying the results. It can be compared against a golden reference simulator to discover most bugs. A functional test generator can also help the verification of the ADL description itself.

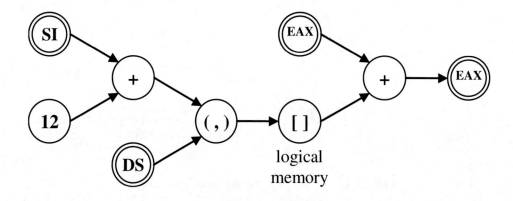

Figure 3. BDF for add

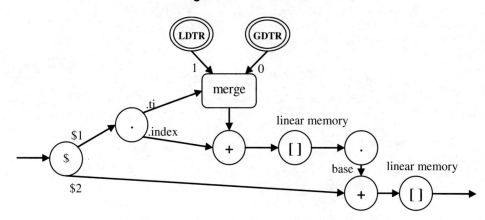

Figure 4. BDF for logic to linear translation

5 Conclusions

In this paper we presented an ADL custom-designed for the IA32 architecture. It introduces several novel features to enable analyzable and executable description of the special features of IA32. We expect to use the ADL to provide an behavioral description of IA32 and its successors for an instruction set simulator and an functional test generator.

6 Acknowledgment

This research is funded by an Intel gift grant in the area of automatic functional test generation from ADL.

References

[1] H. Akaboshi. *A Study on Design Support for Computer Architecture Design*. PhD thesis, Department of Information Systems, Kyushu University, Japan, 1996.

[2] S. Amellal and B. Kaminska. Scheduling of a control and data flow graph. pages 1666–1669, 1993.

[3] M. R. Barbacci. Instruction set processor specifications (ISPS): The notation and its applications. *IEEE Transactions on Computers*, C-30(1):24–40, 1 1981.

[4] J. Buck. *Scheduling dynamic dataflow graphs with Bounded Memory Using the Token Flow Model*. PhD thesis, U. C. Berkeley, 1993.

[5] A. Fauth, J. V. Praet, and M. Freericks. Describing instruction set processors using nML. In *Proceedings of Conference on Design Automation and Test in Europe*, pages 503–507, Paris, France, 1995.

[6] Free Software Foundation. Using the GNU Compiler Collection, http://gcc.gnu.org/onlinedocs/gcc-3.4.0/gcc (current July 2005).

[7] G.De.Jong. Dataflow graphs: System specification with the most unrestricted semantics. In *Proceedings of the European Design Automation Conference*, pages 401–407, 1991.

[8] G. Hadjiyiannis, S. Hanono, and S. Devadas. ISDL: An instruction set description language for retargetability. In *Proceedings of Design Automation Conference*, pages 299–302, June 1997.

[9] A. Halambi, P. Grun, V. Ganesh, A. Khare, N. Dutt, and A. Nicolau. EXPRESSION: A language for architecture exploration through compiler/simulator retargetability. In *Pro-*

ceedings of Conference on Design Automation and Test in Europe, pages 485–490, 1999.

[10] Intel Corporation. *IA32 Intel Architecture Software Developer's Manual*, 2005.

[11] G. Kahn. The semantics of a simple language for parallel programming. pages 471–475, 1974.

[12] E. A. Lee and D. G. Messerschmitt. Static scheduling of synchronous data flow programs for digital signal processing. *IEEE Transactions on Computers*, 36(1):24–35, 1987.

[13] S. Pees, A. Hoffmann, V. Zivojnovic, and H. Meyr. LISA – machine description language for cycle-accurate models of programmable DSP architectures. In *Proceedings of Design Automation Conference*, pages 933–938, 1999.

[14] W. Qin, S. Rajagopalan, and S. Malik. A formal concurrency model based architecture description language for synthesis of software development tools. In *Proceedings of the ACM SIGPLAN/SIGBED 2004 Conference on Languages, Compilers, and Tools for Embedded Systems (LCTES'04)*, June 2004.

[15] N. Ramsey and M. Fernandez. The New Jersey Machine-Code Toolkit. In *Proceedings of the 1995 USENIX Technical Conference*, pages 289–302, January 1995.

[16] G. Zimmerman. The MIMOLA design system: A computer-aided processor design method. In *Proceedings of Design Automation Conference*, pages 53–58, June 1979.

Automatic Testbench Generation for Rearchitected Designs

Mark Nodine

Intrinsity, Inc., 11612 Bee Caves Road, Bldg. II, Suite 200, Austin, TX 78738
nodine@intrinsity.com

Abstract

This paper describes a tool for automated testbench generation used to compare a design against a cycle-accurate RTL reference model. The advantage of this tool is that it allows testing of design modules as they become ready, using chip-level test suites or random tests. The tool is able to handle arbitrary encodings of RTL signals to design signals, retiming, and even changes to the module boundaries.

1. Introduction

Intrinsity is a fabless semiconductor firm whose patented $FAST_{14}$™ technology is based on three legs: [1]

1. The use of footed NMOS transistor gates which use domino (precharge/evaluate) logic. Since NMOS transistors are typically 2-3 times faster for the same area than PMOS transistors, using primarily NMOS transistors to implement logic provides an immediate speed and area improvement over CMOS logic, which uses complementary NMOS and PMOS logic.
2. Multi-phased clocks, typically 4 phases per cycle. The principal benefit of multi-phased clocks is that they enable circuits to be immune to small amounts of clock jitter and skew, simplifying the process of timing closure.
3. Multi-valued, 1-of-N (one-hot) encodings for the outputs of gates. The benefits of these encodings are more subtle. (1) They save time by not having to interpret binary encodings for quantities that are inherently multi-valued. (2) Since the wires representing the one-hot values are routed together, twizzling when they switch directions, they have a high degree of immunity to noise and cross-talk, which is important in high-speed circuits.

The bottom line of the technology is that Intrinsity can produce chips that are much faster with potentially smaller areas than can be done with conventional synthesized CMOS logic for the same functionality. Accordingly, part of Intrinsity's business model is to take customers' RTL code and re-implement it in Intrinsity's internal 1-of-N design language (NDL) to use $FAST_{14}$ for applications that require higher performance than could be obtained by synthesizing the original RTL.

Verification of semiconductor chips can be done either through simulation or using formal methods. Simulation verification involves running a set of tests on the design and comparing the results of those tests with a reference model that produces the desired results. The framework that presents tests to the design and the reference model and compares the two results to determine correctness is called a testbench.

A useful side-effect of Intrinsity's business model is that the customer's RTL can serve as a cycle-accurate reference model for the re-implemented design. This observation is obviously true for the top level of the design, but to the extent that the re-implemented design follows the same micro-architecture as the RTL design, the RTL design can also be used to check lower levels in the hierarchy as they become ready prior to integration. Ideally, some degree of re-architecting should be permitted while still allowing the RTL design to be used as a reference model.

This paper discusses a tool called `tbgen` that creates a testbench to compare a re-implemented design with a cycle-accurate RTL reference model using a set of one or more testbench configuration files to establish the correspondence between the two designs.

There are a few issues related to $FAST_{14}$ that complicate comparing RTL with NDL designs:

1. The original RTL uses a single clock, but the re-implemented design uses a multi-phased clock.
2. The original RTL uses only binary signals, but the re-implemented design encodes these binary signals as multi-valued signals. The design signals can be described as the results of applying an encoding function to the reference signals.

2. Goals

The `tbgen` tool has the following goals:

1. Automate testbench generation for checking one or more design instances against a cycle-accurate RTL reference model.
2. Use a simple specification called a testbench configuration file (TBC file) to describe how the design instance's module interface implements that of the reference model. This specification doubles as documentation for the interface to make sure that the producer and consumer of every signal are compatible.
3. Allow the same TBC file to be reused independent of the reference model, whether the reference model is a unit, a chip-level, a chip with glue logic around it, or multiple chips.
4. Allow the same TBC file to be reused recursively when a design instance occurs as a hierarchical subinstance within another design instance, so the top-level design

can have its subinstances checked all the way down the hierarchy. Thus, generated testbenches have great visibility into the design: an error is reported for the first checked block producing an incorrect output rather than waiting for that error to propagate to a primary output.

5. Generate testbenches with multiple design instances wired together before the next level up in the hierarchy has been coded.
6. Allow design modules to comprise more than one reference module or to have design modules that smear the boundaries relative to their respective reference modules.
7. Allow design modules to be retimed relative to their reference modules.
8. Automatically convert reference signals to the proper phase for their design signals and, as much as possible, infer the encoding functions without requiring user intervention. Inferring of encodings is possible by comparing the bit width of the reference signal with that of the corresponding design signal.
9. Provide useful testbenches before a design is complete.

2.1 Implicit defines

An important design goal for `tbgen` is that a TBC file should be independent of the location of any reference instance within the hierarchy of the reference model. This characteristic allows use of the same TBC file for a unit, block, full-chip, or even multiprocessor testbench as well as to specify checking when the design instance is not the top-level TBC file in the testbench. To this end, `tbgen` produces an implicit define for every module in the reference design that refers to its instance, as long as the module has a unique instantiation in the hierarchy of the reference model. This implicit define acts like a `define` for the module name pointing to the instance. If there is more than one instance of a given module in the reference model, an implicit define can be specified on the command line using the `-D` flag for a top-level TBC file or in the `subinst` statement for a subinstance TBC file.

3. Example Testbenches

This section gives examples of testbenches that can easily be created using `tbgen`. In every case, the testbench port list matches that of whatever reference model is used so the testbench is plug-compatible with the reference model.

Unit-level vs. unit-level. The simplest testbench is when a unit design block is tested in lockstep with a unit-level reference model, assuming that unit-level stimulus is available. Such a testbench is shown in Figure 1 if b is a design re-implementation of RTL block B. This testbench captures the inputs to B to supply them to b and sends the corresponding outputs of b and B to a check. A more detailed representation of this testbench appears in Figure 7.

The advantage of using a unit/unit testbench is that it makes it easier to stimulate the corner cases of the unit, though development of unit-level stimulus can be costly.

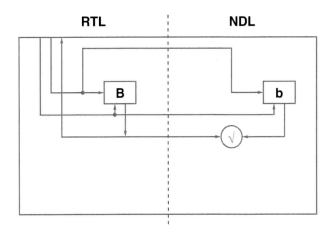

Figure 1. Testing a unit-level design block against a unit-level reference model

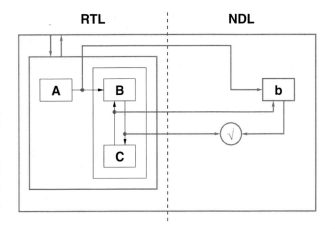

Figure 2. Testing a unit-level design block against a chip-level reference model

Unit-level vs. chip-level. A unit-level design block can be checked against its counterpart while simulating the reference design for the entire chip using the testbench shown in Figure 2. It is like that of Figure 1 except that reference connections of b come from internal nets within the RTL design.

Checking against a hobbled reference model. Sometimes it is important to start checking before all the features of a design block are implemented. It is always possible to generate a standard testbench and restrict the tests to avoid the unimplemented features. However, sometimes the unimplemented features can be disabled in the reference model by tying one or more inputs to a constant value. In that case, we can check the design block against a "hobbled" reference instance, which is a reference instance with one or more inputs tied to constant values to disable features not yet implemented in the design module. The advantage of this approach is that the test space needs no constraining since the hobbled reference model operates as though the test suite did not use the feature. Figure 3 gives an example of a hobbled testbench.

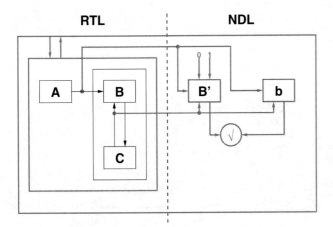

Figure 3. Testing a unit-level design block against a hobbled reference model

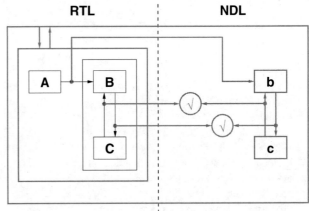

Figure 5. Testing connected design blocks against a reference model

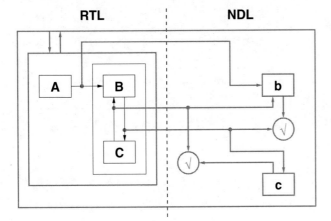

Figure 4. Testing multiple unit-level design blocks against a reference model

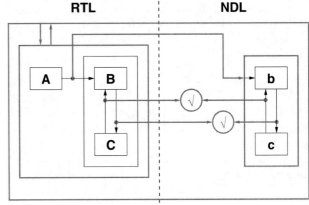

Figure 6. Testing subinstances of a design block against a reference model

Checking multiple design instances independently. Once several design modules are passing their regression tests, it may be useful to put together a testbench where all the design modules are checked independently against their respective modules in the reference module, as shown in Figure 4.

The advantages of such a testbench are:

1. Only one copy of the reference model needs to be simulated in order to check all the design instances.
2. An error in one design instance cannot propagate to cause errors in other design instances since they are not directly connected.

Checking tbgen-connected design instances. It is useful to have tbgen create testbenches where the inputs of one design module are directly wired to the outputs of another design module as shown in Figure 5. This testbench checks compatibility between two modules more stringently than the one in Figure 4 and may be used when the design module that instantiates b and c contains too much unfinished logic to test directly.

Checking subinstances within a design instance. Every chip project eventually enters an integration phase where higher levels of the hierarchy are defined by wiring together lower levels using subinstances. Keeping the output checks for all the instantiated blocks and their instances all the way down the hierarchy makes debugging easier by flagging an error at the time the first checked block produces an incorrect output rather than waiting for that error to propagate to one of the primary outputs. Thus, we want a testbench like that in Figure 6, which for simplicity only shows two levels of hierarchy.

4. Command Line

Here is the command-line usage of tbgen:

`tbgen [options] outfile ref-model cfg-file(s)`

Some options:

-D <*modname*>=<*path*>

    Specify the instance path to use for resolving

`modname
-I <*dir*>
   Search <dir> for .tbc and #include files
-y <*dir*>
   Search <dir> for .v and .vndl files

Each of these options may occur multiple times. Additionally, Verilog options such as +incdir+<dir> are passed to tbgen's internal Verilog preprocessor.

5. Format of a Testbench Configuration File

These are types of statements that can validly appear in a TBC file:
- verbatim statements
- instance definition statements.

In addition, a comment within a TBC file begins with "//" and goes until the end of the line. Whitespace is not otherwise important.

A TBC file can also contain an include directive to include the contents of another file into the input stream, e.g.,

```
#include "<file-name>"
```

5.1 Verbatim statements

A verbatim statement specifies Verilog code to be included verbatim into the testbench. There are four verbatim statements, prolog, header, footer, and epilog, whose format is exemplified by the following prolog statement:

```
prolog [flags] {
  <prolog-code>
} [;]
```

The verbatim statements differ only in where the code winds up in the output testbench file as shown in Section 6.1.

If there are multiple verbatim statements for any given section, the code gets concatenated. The prolog code and epilog code should contain only Verilog constructs that are valid outside a module definition. For example, the prolog code can define any Verilog preprocessor constants or functions, set the timescale, or define special-purpose modules. The header code can be used to define wires and registers to be referenced by the automatically-generated code and the footer code. The footer code can provide assignments for the wires and registers defined in the header code. The prolog and header statements should occur prior to any instance definition statement that references a wire or function defined in it.

Verbatim statements accept the following flag:

-recursive
   Include the code even if this file is describing a subinstance in a testbench for a higher-level module.

Implicit defines within a verbatim statement are substituted prior to writing the testbench.

5.2 Instance definition statements

An instance definition statement is either

```
inst <mod-name> [<inst-name>] ( <port-list> ) ;
```
or
```
inst <mod-name> [<inst-name>] ( <port-list> ) {
  body
} [;]
```

The tbgen tool searches for a Verilog file defining module mod-name and expects an instance of it within the testbench with name inst-name. If inst-name is not specified, it defaults to being the same as mod-name.

For a top-level TBC file (a TBC file specified on the command line), an instance definition statement results in an instance being created in the testbench (a testbench instance). The module can be either a design module to create a design instance to act as a design under test (DUT) or a module from the reference model (for example, in creating hobbled reference models) against which the DUT can be checked.

The contents of multiple instance definitions with the same inst-name and same mod-name are treated as if they are concatenated into a single instance definition. This feature may be needed to prevent a forward reference if two instances need to be wired to each other in the testbench. It is an error to have multiple instance definitions using the same inst-name but with a different mod-name.

Port lists. The purpose of the port-list is to specify options and provide a reference connection for each port of the instance. If a port has information specified in more than one line of the TBC file (e.g., if there are multiple port specifications) the final value for any option/connection is used in generating the testbench.

The semantics of a connection are:

1. For an input port,
   a. For a top-level TBC file, the connection is wired into the port of the testbench instance.
   b. For a subinst TBC file, the connection is ignored, since the input port should already be connected in its parent module.
2. For an output port, the connection specifies the expression against which the output port is checked.

A port list is a comma-separated (or comma-terminated) list of port descriptors, where each port descriptor is one of

```
.<port-reg-exp>  ([<port-flags>] [<verilog-exp>])
match            (<instance-exp>)
feed             (<instance-name>)
```

Explicit port declarations. The first form, referred to as an explicit port declaration, looks similar to a Verilog named port connection, but is different in that <port-reg-exp> is a regular expression and the expression in parentheses may contain port flags.

Some important port flags are:

-delay <*integer*>
   The number of full clock cycles to delay the reference signal before feeding into a design input or checking against a design output. It can be negative,

if the design input/output precedes that of the reference. The reference signal is always delayed to the appropriate phase of the clock cycle, so the -delay flag is needed only if the signal is produced outside the home clock cycle.

-encode <*function-name*>
　The name of a function used to encode the reference signal to the same radix as the design input/output.

-mask <*verilog-expr*>
　A one-bit Verilog expression or constant that specifies when checking should be done; 0 means never do checking. Applies to **outputs only**, default is 1.

-valid <*verilog-expr*>
　A one-bit Verilog expression or constant that specifies when the signal is valid. It is passed as a second argument to the encoding function, which should return whatever the design code is expected to produce for the signal when it is invalid.

An explicit port declaration may also specify a connection for the port(s) matched by the regular expression. This connection can be any Verilog expression, and may or may not have the same width as the matched port; an encoding function is used if the widths differ. The Verilog expression should refer only to instances within the reference model using implicit defines, described above. The regular expression can use parentheses to capture substrings; a connection can contain $1, $2, etc., to substitute these captured values.

For example, an explicit port declaration
　.byp_def_B(.+)_(.+) ()
matches

Signal	$1	$2
byp_def_B31_30	'31'	'30'
byp_def_B29_28	'29'	'28'
...		
byp_def_B1_0	'1'	'0'

Match functions. The match function specifies an instance, which is usually an instance within the reference model specified with an implicit define, but can also be the name of an instance specified earlier in the TBC file. The purpose of the match function is to match any design ports that have not yet specified a connection with those of the instance that can be inferred to represent the same signal by following some naming conventions.

Feed functions. The feed function specifies an instance, which is the name of an instance specified earlier in the TBC file. Using it causes any output ports of the named instance to be connected to any identically-named, unconnected ports of the testbench instance being defined.

Encoding functions. In order to drive RTL signals into inputs of a design instance or to compare RTL signals against the outputs of a design instance, it is necessary to use a common encoding. The tbgen tool is set up to require only functions for encoding RTL signals into the design signals, and not to require the corresponding decoding functions.

The tbgen tool uses these rules to infer an encoding function converting a reference signal to a design signal:

1. If there is a function specified with the -encode flag in an explicit port declaration for the port, use it ELSE
2. If the reference signal and design signal have the same number of bits, use the null encoding ELSE
3. Map reference to design using standard $1 \rightarrow 2$, $2 \rightarrow 4$, and $3 \rightarrow 8$ bit conversions, if applicable, ELSE
4. Report an error.

An error is also reported if the width of an -encode function does not equal that of the port.

Instance statement body. The body of an instance statement contains zero or more of the following statements:

```
subinst [<flags>] <inst-name> [<tbc-name>] ;
check .<sig-reg-exp> ([<flags>] verilog-exp);
```

Subinst statements. The subinst statement specifies the name of an instance reachable within the current module for which checks should be instantiated recursively in the testbench. There must be a TBC file available for the module of which inst-name is an instance, unless tbc-name is specified, in which case it is used as the module name instead. The subinst statement can have the following flag:

-D <*modname*>=<*relative-path*>
　Specify an implicit define override, where modname is the name of a reference module and relative-path gives the value that the implicit define for that module will have during the parsing of the subinstance's TBC file, starting from an existing implicit define. No -D options are needed if the subinstance's TBC file does not use an implicit define for any ambiguous modules.

The processing of a TBC file is a little different when it occurs as the result of a subinst statement. In particular,

1. No instances are created in the testbench for any instance definition statements in the TBC file; it is assumed that they are already instantiated and wired by the current module's definition.
2. Verbatim statements are ignored unless they have the -recursive flag.

The net effect of a subinst statement is to include in the testbench checks for the outputs, internal signals, and subinstances mentioned by its associated TBC file.

Check statements. The check statement specifies a check for an internal net. It looks just like an explicit port declaration except that the sig-reg-exp can match any signal (input, output, reg, wire) within the testbench instance being defined. It accepts the same port flags as any output port in an explicit port declaration.

For example, to create a check for an internal net in mymod against an internal net in `refmod using the encoding function ternary_encode, write:

```
inst mymod (
    match (`refmod)
) {
    check .my_internal_net_B(.*)
          (`refmod.internal_net[$1]
           -encode=ternary_encode)
};
```

6. Processing Done by `tbgen`

Here is the processing done by `tbgen` when it is invoked:
1. It reads the Verilog file defining the reference model and recursively reads the Verilog files for all its subinstances.
2. It computes paths for every unique module within the reference model for use in the implicit defines.
3. It creates an instance of the top-level reference module within the testbench and defines the testbench port list to conform to that of the reference top-level reference module.
4. It parses the TBC files.
5. It wires together the testbench.
6. If there were no errors, it writes out the testbench.

6.1 Format of generated testbenches

An output testbench file contains a module definition with verbatim statements placed in their respective locations. The file has the following format:

```
<prolog-code>
module <module-name> ( <port-list> ) ;
<output declarations>
<input declarations>
<wire declarations>
<reg declarations>
<header-code>
<wire assignments>
<reg assignments> (in always blocks)
<instances>
<checks>
<footer-code>
endmodule
<epilog-code>
```

If inputs are driven according to inputs of a reference block (i.e., for instances in a top-level TBC file), then registers are used that are clocked with the appropriate phase clock. Outputs are always connected to wires (if they connect to anything; they may be unconnected if they are not checked and do not feed another design instance).

The encoding functions and registers are shown schematically in Figure 7, where boxes labeled f represent encoding functions. The flops are clocked with the phase appropriate to the signal (denoted ϕ) and the check is clocked with a slightly delayed version of the phase clock (ϕ').

6.2 Implementation of checks

Checks are implemented by converting the reference signal to a design signal using the appropriate encoding function and clocking it to the same phase as that of the design's output port or internal net that is to be checked. The two signals to be compared are then passed into an `always` block clocked with a slightly-delayed version of the phase clock (to avoid race conditions), as shown in Figure 7.

Error checking uses the following algorithm:
1. If the signal's mask is 0, return OK, ELSE

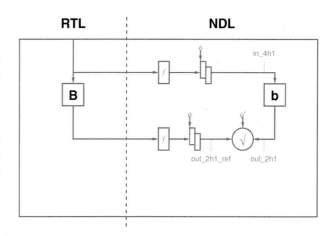

Figure 7. Default register clocking within a testbench

2. If the reference signal is x or z, the signal is considered to be a don't-care at this time and so no checking is done; return OK, ELSE
3. If the reference signal equals the design signal, return OK, ELSE
4. Return NOT_OK.

6.3 Effect of delays on testbenches

When retiming occurs, `tbgen` adds additional flops to delay signals to their appropriate clock. There are four retiming cases to be considered: (1) positive delay on an output signal, (2) negative delay on an output signal, (3) positive delay on an input signal, and (4) negative delay on an input signal. Of these cases, the first three are easy to deal with and the fourth is more difficult.

The signal names for delayed signals within the testbench are the same as the original names, except that they have as a suffix +<n> for a signal delayed n cycles relative to the original signal or a suffix of -<n> for signals that are accelerated n cycles relative to the original signal. The frame of reference for all the signals is with respect to the undelayed cycle for the design module. Since these suffixes contain characters (+ and -) that are not normally valid in a Verilog signal name, the names are backslash-quoted.

A positive delay for an output can be accomplished by passing the reference signal through extra flops prior to checking. A negative delay for output is done by instead passing the design output signal through flops prior to checking. A positive delay on an input results in extra flops on the reference signal before entering the design.

Figure 8 shows these retimings schematically. In this figure, `in_4h1` has been connected with `-delay=1` in its port list; accordingly, the encoded/phased reference signal `in_4h1` gets delayed by a flop to produce `in_4h1+1`, which enters the design. Output `o1_2h1` has also specified `-delay=1` in its port list, so the encoded/phased reference signal `o1_2h1_ref` gets delayed to produce `o1_2h1_ref+1`,

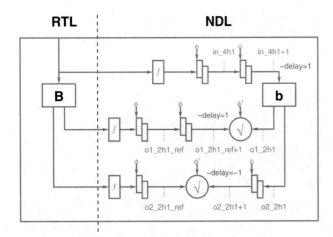

Figure 8. Clocking with delays on outputs and positive delays on inputs

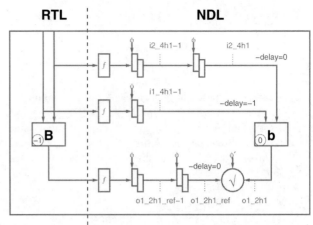

Figure 9. Clocking with a negative delay on an input

which is then compared against `o1_2h1`. Output `o2_2h1` has specified `-delay=-1` in its port list. Since it is not possible to go one cycle into the future to see what the reference model will produce for its signal, the testbench instead delays `o2_2h1` by one cycle to get `o2_2h1+1`, which it then compares against `o2_2h1_ref`, resulting in an effective delay of -1, but checking a cycle later.

As mentioned above, the difficult case is when an input signal has a negative delay. Since it is not possible to go into the future to find out what the input will be, tbgen compensates by shifting all other signals out by the maximum negative delay, resulting in a design that operates in a shifted time zone, or, more precisely, a reference model that operates in a shifted time zone, since the "home" time zone is considered to be that of the design module.

Figure 9 shows an example of a testbench resulting from a single negatively-delayed input with all other inputs and outputs having no delays. (Other delayed signals are adjusted relative to this scheme.) In this example, `i1_4h1` specifies `-delay=-1` and all other signals are undelayed (`-delay=0`). Because of the negative delay, the reference model is operating in clock "-1" relative to the design, so all of its encoded/phased signals start out with a suffix of `-1`. In the case of `i1_4h1`, the `-1` signal is wired directly to the design, but an undelayed input like `i2_4h1-1` is delayed for a cycle to produce `i2_4h1`, which then enters the design. Likewise, the undelayed output `o1_2h1_ref`, which is initially produced in the "-1" clock as `o1_2h1_ref-1` is delayed for a cycle to produce `o1_2h1_ref`, which is checked against `o1_2h1`. Each instance is annotated with a circle surrounding a number that specifies its time zone.

7. Example Testbench Specifications

This section contains examples of some common kinds of testbench specifications.

7.1 Simplest testbench

In the simplest case of a module that instances no other modules and follows the naming conventions assumed by the `match` function, a TBC file can be as simple as

```
inst mymod (
    #include ``commonports.tbh''
    match(`refmod)
);
```

In this and all ensuing examples, it is assumed that any include files are somewhere on the path specified by the `-I` command-line flag. The file `commonports.tbh` contains an explicit port declaration for each non-functional (e.g., scan) port.

The TBC file above says to instantiate an instance of module `mymod` named `mymod` (since no instance name was supplied) and to create a port correspondence with the instance of some reference module `refmod`. This reference instance can either be unique within the reference model or it can be specified on the command line with `-Drefmod=<dotted-path>`.

If a design follows the naming conventions, a TBC file like that above can be used to create a testbench like either that of Figure 1 or Figure 2 by invoking tbgen as follows:

```
tbgen unit.v B b   # Unit-level ref. model B
tbgen chip.v TOP b # Chip-level ref. model TOP
```

7.2 Testbench with exceptions

Here is a TBC file for a module which has some explicit exceptions to the defaults:

```
inst mymod (
  .funny_name        (`refmod.StandardName),
  .bus_B(.+)_(.+)    (`refmod.SomeBus[$1:$2]),
  .delayed_sig       (-delay=1),
  .unchecked_out     (-mask=0),
  .weird_encode      (-encode=myfunc),
  .encoded_out       (`refmod.EncodedOut ?
                       2'b01 : 2'b10),
  .inverted_in       (~ `refmod.InvertedIn),
  match              (`refmod)
```

```
);
```
This example has the following exceptions in its port list:

`funny_name`
> Can't be matched automatically, so gives an explicit connection. Note that `StandardName` is probably the name of a port on the reference instance, but it could also be an internal wire or reg.

`bus_B(.+)_(.+)`
> Can't be matched automatically, so gives an explicit connection for sub-bits within the bus. Each signal corresponds to a pair of bits in the original. Thus, `bus_B31_30` is connected to `SomeBus[31:30]`, etc.

`delayed_sig`
> Follows naming conventions, so no connection expression is needed, but the testbench instance is delayed by one cycle from the reference instance on this port.

`unchecked_out`
> Specifies not to check a matched output.

`weird_encode`
> Specifies an explicit encoding function.

`encoded_out`
> Specifies an explicit encoding of the 1-bit reference signal `InvertedOut` into a 2-bit value.

`inverted_in`
> Specifies that the 1-bit reference signal `InvertedIn` should be inverted before being passed to the implicit $1 \rightarrow 2$ bit encoding function.

Note that it doesn't matter whether the `match` function comes before or after the explicit port declarations, since it provides connections only for unconnected ports, whereas a connection in an explicit port declaration overrides any previous connection.

7.3 Hobbled reference models

Figure 3 showed a testbench that uses as its reference a hobbled instance of the reference module in which some inputs are tied off. The following shows how such a testbench can be created at its simplest:
```
inst B B' (
    match          ('B),
    .TiedOffInput  (1'b0),
);
inst mymod (
    match          (B'),
);
```
In this example, we create a testbench instance of reference module B called B' (not actually a legal name) which corresponds exactly to the instance in the reference model with the exception that port `TiedOffInput` is held at 0. We now specify that our design module `mymod` be instantiated with inputs and outputs that correspond to those of B'.

If none of the design blocks has an input with a negative delay, then the above TBC file will produce a testbench like

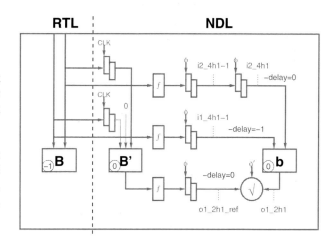

Figure 10. Testbench with a shifted hobbled reference instance

that of Figure 3. However, in the event that some design block has an input with a negative delay, the above will produce a testbench where the hobbled model is operating in the same clock as the reference model and off by one or more cycles from the design modules. For ease of debugging, though, it is possible to specify a time-shifted hobbled model in which the hobbled RTL model is shifted into the same time zone as that of the design module. This modified testbench is shown schematically in Figure 10.

A variant of creating a hobbled instance of a reference module is to create a debug instance which operates exactly the same as the reference instance but has been shifted into the same clock cycle as the design.

7.4 Independent design instances

It is very simple to create a testbench with more than one design instance in it where the instances are not connected to each other, such as that of Figure 4: simply specify each of the design module names on the command line. For example, to create a testbench with instances of both modules `mymod1` and `mymod2`, create the testbench using
```
tbgen output.v refmod mymod1 mymod2
```

7.5 Dependent design instances

To speed up integration, it is important to be able to create a testbench with more than one design instance where the outputs of one are directly wired to the inputs of another as shown in Figure 5.

If the outputs of one module have identical names to the inputs of the second, the `feed` function can be used to automate the wiring as follows:
```
inst mymod1 (
    match('refmod1)
);
inst mymod2 (
```

```
    feed           (mymod1),
    .unmatched_in  (mymod1.some_out),
    .inverted_in   (~ mymod1.inverted_in),
    match          (`refmod2)
);
```

Here the `feed` function wires output ports of `mymod1` to input ports of `mymod2` with the same name. There is an explicit port declaration for port `unmatched_in` since the names do not match, and for `inverted_in`, since even though the port name matches, the signal needs to be inverted. Any other inputs and all outputs are matched to the reference instance referred to through `` `refmod2``. It is important for the `feed` function to precede the `match` function, since `feed` does not have any effect for a port whose connection is already specified.

If two design instances need to be wired to each other, it is necessary to use an incomplete instance statement for one instance to avoid having a forward reference, as shown in the following example:

```
inst mymod1 () ;    // Declare mymod1
inst mymod2 (
    feed (mymod1),
    match(`refmod2)
);
inst mymod1 (       // Define mymod1
    feed (mymod2),
    match(`refmod1)
);
```

7.6 Checking subinstance outputs

If a design module instantiates another design model for which a TBC file is available, it is possible to re-use that TBC file to create checks for the outputs of the subinstance using the `subinst` statement in its body; the subinstance itself may in turn specify its own subinstances for checking. The resulting testbench looks like that of Figure 6.

The following example shows how subinstance checking might be coded:

```
inst mymod1 (
    match(`refmod1)
) {
    subinst myinst;
};
```

It is an error if the module `mymod1` does not contain an instance whose name is `myinst`. If `myinst` is an instance of module `mymod2`, then `tbgen` searches the path specified using the `-I` command-line flag for `mymod2.tbc`, the testbench configuration for `mymod2`. As long as there is only one instance of `refmod2` in the reference model, this scheme is sufficient. If multiple instances of `refmod2` exist, the `subinst` statement needs to specify the reference instance that corresponds to it:

```
inst mymod1 (
    match(`refmod1)
) {
    subinst myinst1 refmod2=`refmod1.refinst1;
    subinst myinst2 refmod2=`refmod1.refinst2;
};
```

This extra argument has the effect of defining `` `refmod2`` to `` `refmod1.refinst1`` and `` `refmod1.refinst2`` for the two subinstances while processing `mymod2.tbc`.

7.7 Checking internal signals

Sometimes a designer knows that an internal signal in a design instance corresponds to an internal signal (or combination of internal signals) in the reference instance and wants to know immediately if the design signal's value fails to match. A common example might be an architected register file. It is possible to specify extra checks within a TBC file as shown:

```
inst mymod (
    match(`refmod)
) {
    check .mysig_B(.+)
        (-delay=1 `refmod.Internal_Sig[$1]);
};
```

8. Results

Intrinsity has adopted a methodology based on `tbgen` to verify the implementation of a high-speed core. This methodology has been very successful in allowing early testing of the design modules prior to integration. Furthermore, the ability to specify the interface of the design modules in terms of the RTL reference model, both with respect to timing and to re-architecting the module boundaries, has the following benefits:

1. It provides good documentation of those interfaces.
2. It ensures the consumers of a signal are driven the same way that it checks the producer of that signal, which reduces the possibility of independent designers making different assumptions about the signal.
3. The fact that interface changes are so straightforward to specify has freed designers to make needed changes to the micro-architecture to accommodate a high-speed implementation without compromising the ability to do early testing.

References

[1] S. Horne, D. Glowka, S. McMahon, P. Nixon, M. Seningen, and G. Vijayan, "Fast$_{14}$ technology: Design technology for the automation of multi-gigahertz digital logic," in *International Conference on Integrated Circuit Design and Technology*, pp. 165–173, 2004.

Author Index

Ahuja, Sumit	3	Mathaikutty, Deepak A.	3
Al-Sukhni, Hassan	8	Mauri, Robert	85
Bauserman, Adam	91	Myers, Chris	109
Becker, Bernd	33	Nodine, Mark	128
Ben-Tzur, Asa	119	Palma, W. Di	77
Bertacco, Valeria	91	Pravadelli, Graziano	98
Bojan, Tommy	85	Qin, Wei	119
Chakraborty, Supratik	51	Ravotto, D.	71, 77
Chiou, Derek	8	Ray, Sandip	25
DeOrio, Andrew	91	Reorda, M. Sonza	71, 77
Dingankar, Ajit	3	Roy, Subir K.	63
Frumkin, Igor	85	Sanchez, E.	71, 77
Fummi, Franco	98	Schillaci, M.	71, 77
Gutkovich, Boris	119	Sen, Alper	109, 44
Harris, Ian G.	98	Shukla, Sandeep	3
Herbstritt, Marc	33	Shukla, Sandeep K.	39
Holt, Jim	8	Singh, Gaurav	39
Hunt Jr., Warren A.	25	Squillero, G.	71, 77
Ikiz, Selma	44	Struve, Vanessa	33
Lee, Lung-Jen	15	Sunkari, Sasidhar	51
Little, Scott	109	Sunwoo, Dam	8
Maneparambil, Kailasnath	51	Tseng, Wang-Dauh	15
Marconcini, Cristina	98	Vedula, Vivekananda	51

IEEE Computer Society Conference Publications Operations Committee

CPOC Chair
Chita R. Das
Professor, Penn State University

Board Members
Mike Hinchey, *Director, Software Engineering Lab, NASA Goddard*
Paolo Montuschi, *Professor, Politecnico di Torino*
Jeffrey Voas, *Director, Systems Assurance Technologies, SAIC*
Suzanne A. Wagner, *Manager, Conference Business Operations*
Wenping Wang, *Associate Professor, University of Hong Kong*

IEEE Computer Society Executive Staff
Angela Burgess, *Executive Director*
Alicia Stickley, *Senior Manager, Publishing Services*
Thomas Baldwin, *Senior Manager, Meetings & Conferences*

IEEE Computer Society Publications
The world-renowned IEEE Computer Society publishes, promotes, and distributes a wide variety of authoritative computer science and engineering texts. These books are available from most retail outlets. Visit the CS Store at *http://www.computer.org/portal/site/store/index.jsp* for a list of products.

IEEE Computer Society *Conference Publishing Services* (CPS)
The IEEE Computer Society produces conference publications for more than 250 acclaimed international conferences each year in a variety of formats, including books, CD-ROMs, USB Drives, and on-line publications. For information about the IEEE Computer Society's *Conference Publishing Services* (CPS), please e-mail: cps@computer.org or telephone +1-714-821-8380. Fax +1-714-761-1784. Additional information about *Conference Publishing Services* (CPS) can be accessed from our web site at: *http://www.computer.org/cps*

IEEE Computer Society / Wiley Partnership
The IEEE Computer Society and Wiley partnership allows the CS Press *Authored Book* program to produce a number of exciting new titles in areas of computer science and engineering with a special focus on software engineering. IEEE Computer Society members continue to receive a 15% discount on these titles when purchased through Wiley or at: *http://wiley.com/ieeecs*. To submit questions about the program or send proposals, please e-mail jwilson@computer.org or telephone +1-714-816-2112. Additional information regarding the Computer Society's authored book program can also be accessed from our web site at:
http://www.computer.org/portal/pages/ieeecs/publications/books/about.html

Revised: 21 January 2008

CPS Online is our innovative online collaborative conference publishing system designed to speed the delivery of price quotations and provide conferences with real-time access to all of a project's publication materials during production, including the final papers. The *CPS Online* workspace gives a conference the opportunity to upload files through any Web browser, check status and scheduling on their project, make changes to the Table of Contents and Front Matter, approve editorial changes and proofs, and communicate with their CPS editor through discussion forums, chat tools, commenting tools and e-mail.

The following is the URL link to the *CPS Online* Publishing Inquiry Form:
http://www.ieeeconfpublishing.org/cpir/inquiry/cps_inquiry.html